CONTRIBUTIONS FROM THE MUSEUM OF GEOLOGY
UNIVERSITY OF MICHIGAN

VOLUME I

THE STRATIGRAPHY AND FAUNA OF THE HACKBERRY STAGE OF THE UPPER DEVONIAN

THE MACMILLAN COMPANY
NEW YORK   BOSTON   CHICAGO
ATLANTA   SAN FRANCISCO

MACMILLAN & CO., Limited
LONDON   BOMBAY   CALCUTTA
MELBOURNE

THE MACMILLAN CO. OF CANADA, Ltd.
TORONTO

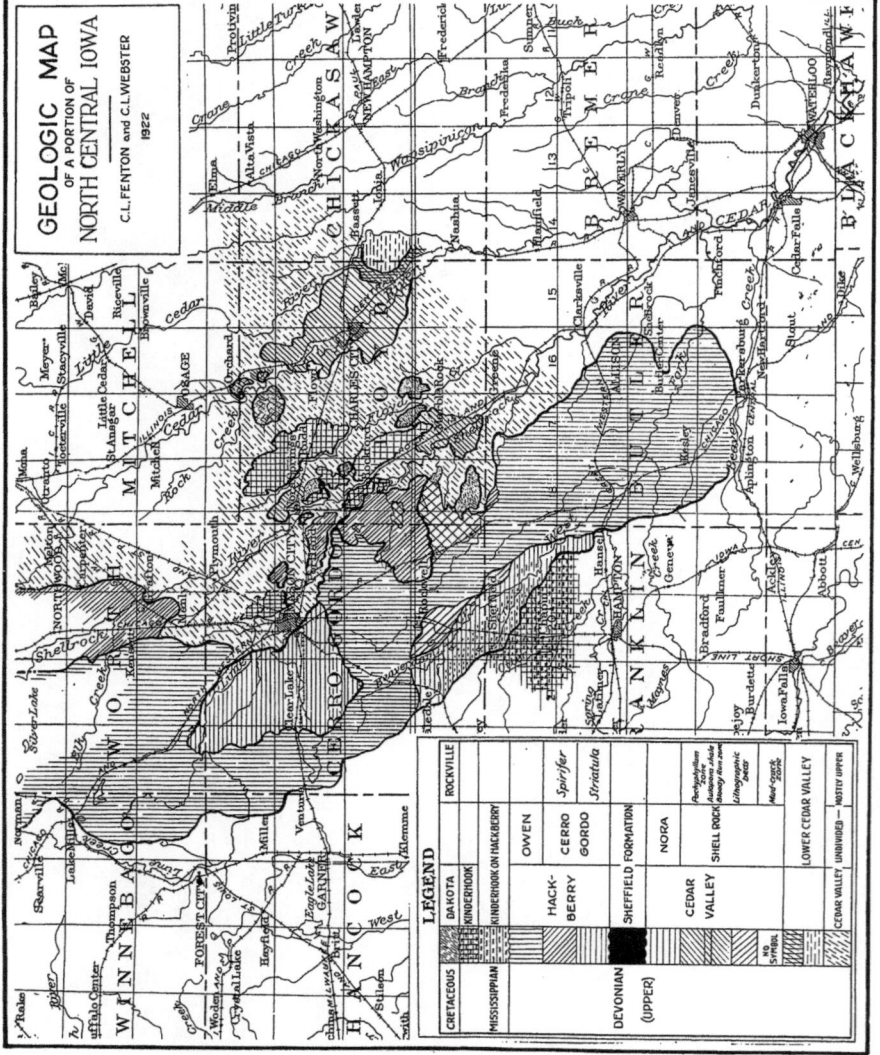

MAP SHOWING THE HACKBERRY STAGE AND ASSOCIATED FORMATIONS

THE STRATIGRAPHY AND FAUNA
OF THE HACKBERRY STAGE
OF THE UPPER DEVONIAN

BY
CARROLL LANE FENTON
AND
MILDRED ADAMS FENTON

New York
THE MACMILLAN COMPANY
LONDON: MACMILLAN & COMPANY, Limited
1924

PREFACE

THE rocks of the Hackberry stage are exposed throughout a narrow belt in north central Iowa, where they constitute the uppermost member of the Devonian section. Although limited in area and in thickness, the Hackberry contains an abundant fauna, preserved with unusual excellence. A number of the species have been described and illustrated in the publications of Hall, Hall and Whitfield, Hall and Clarke, Calvin, Webster and others, but most of them have remained undescribed or erroneously identified with eastern species. Our purpose is to furnish a detailed account of the Hackberry strata, with sections, accompanied by an adequate — though not complete — discussion of the fauna. In the fauna, most of the typical species are described, as well as many rarer ones. In a few cases, detailed treatment of varieties, evolution and association has been possible. Lack of time has prevented description of the Protozoa and the Stromatoporoidea, of which there are numerous species, most of which are undescribed.

Some points in nomenclature may deserve comment. We have used 'stage' in much the same sense that the United States Geological Survey uses 'formation.' 'Zone' is used to indicate a unit that is distinguished both faunally and stratigraphically. Thus it is an easy matter to distinguish, either by fossils or by rocks, the Spirifer zone, or the Idiostroma zone. 'Faunule,' on the other hand, indicates a biologic group, an association of animals, living at approximately the same time and in the same region, and more or less distinct from other groups, that may or may not have been contemporary. Thus

vii

the Fucoid faunule of the Bird Hill locality is lacking at Hackberry Grove, where it is replaced by the Brachiopod faunule. In some cases, notably the Leptostrophia faunule, the stratum is distinct, but generally it is not, so that the faunules have little stratigraphic significance.

We have been greatly aided by various people. Mr. Clement L. Webster, of Charles City, Iowa, who has lent many rare specimens for study, has supplied notes, and even cuts illustrating specimens that were not available for study and illustration. It is only fair to Mr. Webster to say that difficulty of communication has prevented consultation, so that, with the exception of new species, our opinions do not of necessity represent his. Had his collection been available for study, we should, perhaps, have retained a number of species that we have felt obliged to abandon as being insufficiently described. Others who have assisted are Mr. G. M. Ehlers, Mr. A. W. Slocom, Dr. E. M. Kindle, and Dr. Stuart Weller. In addition to giving opinions on various troublesome questions, Dr. Weller has read the entire manuscript.

C. L. F.
M. A. F.

ANN ARBOR, MICHIGAN
June, 1923

CONTENTS

ILLUSTRATIONS

Map showing the Hackberry Stage and associated

LIST OF FIGURES IN THE TEXT

I. STRATIGRAPHY

IN the northeast quarter of section 35, Portland Township, Cerro Gordo County, Iowa, lies Hackberry Grove, the type locality of the principal formation described in this paper. On the right (south) bank of Lime Creek at this point is an escarpment which has an average height of about 75 feet above the normal water level of the creek. At the creek bed, extending upward for about 40 feet is a gentle, talus-covered slope, beneath which lies the clayey shale of the Sheffield formation. Above this shale, which is fine, plastic and of a uniform blue-gray color, lie the blue, blue-brown and yellow shales of the Cerro Gordo division of the Hackberry stage. Above these are six to seven feet of limestones and shaley limestones which constitute the basal member of the upper, or Owen, substage of the Hackberry.

A. THE NAME 'HACKBERRY'

In the case of this, as in many other formations, there is a lack of uniformity in names used; the name 'Lime Creek' commonly appears in place of 'Hackberry,' as well as in place of both 'Hackberry' and 'Sheffield.' Other names, such as 'Rockford Shales,' have been used, as well as the earlier 'Hamilton' and 'Chemung' of Hall and other geologists. The term 'Rockford Shales,' used by Webster in 1887, was preoccupied by 'Rockford Goniatite Beds,' and so had to be abandoned. In 1889 Webster proposed the name 'Hackberry,' to include all beds between the shale to which C. L. Fenton applied the name 'Sheffield' (1919) and the Kinderhook limestones and shales. In 1897 Calvin proposed the name 'Lime Creek,' retain-

1

ing Webster's 'Hackberry' as a subdivision, and greatly chang-
ing the meaning of the term.   Since that time the Iowa Survey
has used Calvin's name, while authors publishing through other
mediums have taken their choice.   In 1919 the senior author of
this paper discussed the confusion of terms, but the fact that
in the very month this is being written there has appeared a
paper in which the name 'Lime Creek' is used seems to justify
a republication of the discussion referred to.

"In 1889 there appeared, in the American Naturalist (vol.
23, pp. 229–243), a paper by Clement L. Webster entitled 'A
General Preliminary Description of the Devonian Rocks of
Iowa; which constitute a typical section of the Devonian
Formation of the Interior Continental Area of North America.'
In this paper both of the formations exposed at Hackberry
Grove were described, that which I have referred to as the
Sheffield being placed as Hamilton in age.   The 'Hackberry
Group' as proposed by Webster in this paper 'is known to
attain a thickness of forty-five feet, and is made up, for the
greater part, of a yellowish brown argillaceous and sometimes
arenaceous, shaley limestone;' this formation also is the highest
division of the rocks of this age [the Devonian] in the state.
Detailed faunal lists were given by Webster, and the formation
was described as fully as the condition of the exposures at that
time would permit.   In later papers Webster gave additional
information regarding this formation.

"In 1897, Calvin (Iowa Geol. Survey, vol. VII) published a
description of what he called the 'Lime Creek Stage' [Shales],
which according to his definition includes not only the Hackberry Stage of this paper (and of Webster), but also the Sheffield
formation.   Calvin applied the name 'Owen Sub-stage' to those
rocks here placed in that division, and 'Hackberry Sub-stage'
to the remainder of the Hackberry and the whole of the Sheffield.
This division of Calvin's was adopted by the Iowa Survey —

of which Calvin was director — and is at present in more general use than is Webster's classification.

"When it is noted that Calvin used 'Hackberry' to designate one of the subdivisions of his 'Lime Creek Stage,' it is naturally questioned whether or not Webster's Hackberry was made to include substantially all of the Owen, and on investigation it will be found that Webster undoubtedly included all of the Devonian rocks overlying the Sheffield formation and underlying the Kinderhook in the district where the Hackberry is known to occur. It is true that he recognized at that time but fifteen feet of the Owen Sub-stage, but there can be no doubt that he recognized at least part of the Owen, and included it in his Hackberry Group. Several of the typical fossils of the Owen were described or mentioned by Webster. . . . [These fossils are *Pachyphyllum crassum, P. crassicostatum, Westernia gigantea, W. gigantea owensis,* and the species *A cervularia bassleri,* which was then considered as a form of *A. davidsoni.*]

"It is plain that whatever weight there may be in priority of publication is on the side of the term 'Hackberry.' But there is another consideration that seems to be even more important, and that is the way in which Calvin's terminology was evolved. The Iowa Survey's process was this: It took a named formation, added to that formation another (which has no connection with it), and applied an entirely new formation name. It then divided the formation derived by this operation into two sub-stages. To one of these it gave a name that is fitting and suitable — the Owen. For the other it took the name originally proposed for the entire formation; cut it down on one side to cover less than half of what it was originally meant to cover; stretched it on the other to include an unrelated formation that it was never intended to include, and thus produced the Hackberry Sub-stage." [1]

[1] *Am. Journ. Sci.,* 4th Ser., vol. 48, pp. 355–358.

There are two arguments against the use of the name 'Hackberry.' The first of them is that the formation was so indefinitely delineated in Webster's original description that Calvin was thoroughly justified in using a new name. But a field study shows that such is hardly the case; Webster's underestimates are proportionately no greater than Calvin's over-estimates. And by his statement that the 'Hackberry Group' included all rocks above the blue ('Hamilton') shale and below the Kinderhook, Webster gave a definition sufficiently clear for general use. If we were to abandon all stratigraphic names that, in original publication, are no more clearly defined than 'Hackberry,' we should lose almost all of our names for periods, series, stages, and so on, and should find ourselves face to face with the task of finding almost a complete new set of names.

The other argument deals with publication, that is, with abundance of publication rather than any question of either clarity or priority. It is quite true the name 'Lime Creek' has been used more often and by a greater number of authors than has 'Hackberry.' But analysis shows that in most of these cases the name has been used because Calvin and his associates of the Iowa Survey used it. In most cases, these associates did not give attention to the derivation of the name, or the reason for its being.[2] In others, it is plain that the author was so unfamiliar with the formation about which he wrote that he could not give a recognizable section of it,[3] even for localities where the beds can be measured with a tape or foot-rule. Thus the weight of the Iowa Survey publications, as well as

[2] Such is the case of Mr. A. O. Thomas, of the University of Iowa and the Iowa Survey, who has assured the authors that he adopted the name 'Lime Creek' without examining the nomenclature, or determining priority. Mr. Thomas's 'Lime Creek' equals Webster's 'Hackberry.'

[3] Notably Beyer's sections (*Ia. Geol. Surv.*, vol. 14, and others). It is impossible to correlate his Rockford section with the actual strata.

of publications of the name 'Lime Creek' by authors who quoted the Iowa Survey, is considerably lessened.

We, therefore, find three reasons for using the name 'Hackberry' instead of 'Lime Creek:'

1. Our understanding of the formation in question is virtually identical with that of Webster when he proposed the 'Hackberry Group,' includes the same beds, and places the same upper and lower limits. In other words, the formation which we describe *is* the Hackberry Group of Webster's original definition.

2. The formation which we describe is *not* identical with the 'Lime Creek' of Calvin, since we separate the Sheffield Formation from the beds above. Nor does it coincide with Calvin's modification of Webster's 'Hackberry,' which includes the Sheffield but excludes the Owen.

3. 'Hackberry' was recognizably defined eight years before Calvin's 'Lime Creek' appeared; therefore it clearly enjoys priority. Granting that Webster's definition was inferior to Calvin's in definiteness, it is still quite recognizable. And we feel that, where there is no strong reason to the contrary, priority should apply to stratigraphic names as it does to biologic ones. Such a reason we have failed to find.

B. The Sheffield Formation

In Webster's paper in the *American Naturalist* he referred to a blue-gray shale formation, apparently non-fossiliferous, which lies below the Hackberry, and is disconformable with it. He did not, however, suggest any name for the shale member. Calvin mentioned this same formation and, as noted, considered it as the base of the 'Lime Creek,' principally, it seems, because of the shaley nature of the beds. In 1908 Webster published a paper entitled *An Interesting Fauna and Flora Discovered*

below the Hackberry Group, at Rockford, Iowa.[4] In this paper
he discusses and illustrates a number of plant and animal fossils
found in this blue shale. These fossils include various plants,
a *Lingula,* and a fragment of an arthropod, probably a trilobite.
A single specimen gives some evidence of being a fragment of a
very small cephalopod.

Further collection in the blue shales has afforded great
numbers of *Lingulas* and quantities of plant remains, some of
which indicate that the actual organisms were of considerable
size, and grew either near shore or in very quiet and shallow
bays and lagoons. None of the fossils have been studied in
detail, so that identifications are not available. A few years
ago we secured from the plant foreman of the Rockford Brick
and Tile Co. some very fine fossils of sponges. These we lent
to Mr. Thomas for study and description.[4a] The striking thing
about these sponges is their great size as compared with the
minute *Lingula* and arthropod. Also, there is a very decided
contrast between the marine animals and the apparently estu-
arine or even terrestrial plant remains.

Turning from paleontology to lithology, we find a marked
difference in the shales of the Hackberry and the formation
underlying it. The former is characterized by very calcareous
beds, commonly indurated to a certain extent, and in the lower
portions very gritty and strongly iron-stained. The latter is
made up of blue-gray, plastic clay-shales, of nearly uniform
texture throughout and with no induration. Chemical analyses
furnished by the Iowa Survey [5] indicate the following to be the
typical composition of the beds:

[4] *Contributions to Science,* vol. 1, pp. 5–8, with Pl. I. This plate is
reproduced as Pl. XLII of this publication. *Contributions to Science,* a private
periodical issued by Webster, is now discontinued.

[4a] See *Some New Paleozoic Glass-sponges from Iowa,* by A. O. Thomas.
Proc. Ia. Acad. Sci., vol. 29, pp. 85–91, Plate. Species described as
Iowaspongia annulata.

[5] *Ia. Geol. Surv.,* vol. 7, p. 191.

Hygroscopic water	.85
Combined water	3.74
CO_2	4.80
SiO_2	54.64
Alumina	14.62
Iron oxide (Fe_2O_3 ?)	5.69
Manganese oxide (MnO ?)	.76
Lime (CaO)	5.16
Magnesia (MgO)	2.90
Soda (NaO)	1.12
Potash (K_2O)	4.77
Total	99.05 %
Error	.95 %

Although accurate analyses of the Hackberry shales are not available, there is no question that the proportions of the lime and iron oxide in them would be greater than in the Sheffield, while the alumina would run lower.

There are, therefore, the following reasons for separating the blue-gray shaley clays of the Sheffield from the shales and limestones of the Hackberry formation:

1. The two formations are separated by a distinct disconformity, which is specially noticeable at Mason City and Rockford.
2. There are striking lithologic differences between the two formations.
3. Their faunas and floras are entirely distinct, not a single species — or even genus — being known to be common to both. Both are of upper Devonian age, but there is no closer relationship between them.

C. STRATIGRAPHY OF THE HACKBERRY STAGE

There is no one point at which a complete section of the Hackberry stage, as it is developed in the state of Iowa, can be secured. Webster, in his paper of 1889, gave a general section, but the exposures available prevented him from securing accurate measurements. Calvin, in 1897, gave a section which, with the exception of the part dealing with the Sheffield Formation, was very accurate. The general section as taken by C. L. Fenton and C. L. Webster, and published by the former,[6] is given below:

HACKBERRY STAGE

	Feet
II. OWEN SUBSTAGE	
5. Acervularia Zone	
Calcareous, light-gray limestones, containing *Pachyphyllum*, *Alveolites* and other corals, with *Acervularia bassleri* as the most distinctive species. Gastropods abundant.......	20
4. Floydia Zone	
Magnesian shales and limestones and argillaceous dolomitic limestones, mostly deep buff or brown. Very fossiliferous, but with the fossils not very well preserved. Numerous variants of the species of *Floydia* the characteristic fossils..	30
3. Idiostroma Zone	
Buff, buff-brown and gray limestones medium to thickly bedded, with some buff shale. Crowded throughout by two species of what appear to be stromatoporoids belonging to the indefinite genus *Idiostroma*. *Floydia* and brachiopods common..	4–6½
I. CERRO GORDO SUBSTAGE	
2. Spirifer Zone	
Yellowish (where unoxidized, bluish) calcareous shales and shaley clays with occasional bands of indurated shale and shaley limestone of buff-gray color. This is the zone of abundant fossils, in which most of the 200 or more species and varieties of the Hackberry fauna are to be found.....	20
1. Striatula Zone	
Calcareous shales and shaley limestones, slightly to very gritty; they contain much pyrite near the base, and in the two lowermost feet, many concretions. In the lower beds fossils are almost exclusively casts, and the range of species is limited. Fucoids (*Gracilerectus*) common throughout; *Gypidula cornuta* common in the uppermost beds.........	12–25

Disconformity

[6] *Am. Journ. Sci.*, 4th Ser., vol. 48, p. 360. *Naticopsis Zone* of that paper is here changed to *Floydia Zone* because of the assignment of *N. gigantea* and its associated species to Webster's genus *Floydia*.

J. Hall 1858	C. A. White 1870	Hall and Whitfield 1873	H. S. Williams 1883	J. M. Clarke 1885	C. L. Webster 1887
HAMILTON	HAMILTON	CHEMUNG	" Rocks of Lime Creek " No forma-tion name proposed, nor definition made.	" Lime Creek Beds " No name proposed, nor definition made.	Rockford Shales (Erroneously includes Cedar Valley strata, particularly near Rudd and Nora Springs, Ia. These strata belong mainly to the Nora For-mation.) Division suggested but not made.

NOMENCLATURE

C. L. Webster 1889	S. Calvin 1897	C. L. Fenton 1918	C. L. Fenton 1919	C. L. & M. A. Fenton 1923
HACKBERRY GROUP — Upper Horizon	LIME CREEK / SHALES — Owen Beds	HACKBERRY STAGE / Owen Substage / (not named) — Acervularia Zone	HACKBERRY STAGE / Owen Substage — Acervularia Zone	HACKBERRY STAGE / Owen Substage — Acervularia Zone
		Naticopsis Zone	Naticopsis Zone	Floydia Zone
		Idiostroma Zone	Idiostroma Zone	Idiostroma Zone
Middle Horizon	Hackberry Beds	Spirifer Zone	Cerro Gordo Substage — Spirifer Zone	Cerro Gordo Substage — Spirifer Zone
Lower Horizon		Striatula Zone	Striatula Zone	Striatula Zone
Genesee		Genesee ??	Sheffield Formation	Sheffield Formation

SHEFFIELD FORMATION

Calcareous clay-shales of very fine and uniform texture.
Fossils absent where clay is weathered. The commonest
species are *Lingula fragilis* Webster and *Iowaspongia annulata* Thomas. The clay is used extensively in manufacture
of brick and tile.................................... 90–100

The two members of the Cerro Gordo substage are highly
variable in thickness, nature of rocks, and in fauna. Thus the
lower member, the Striatula zone,[7] where it is exposed in the
pits of the American Brick and Tile Company at Mason City,
consists of about 25 feet of heavy-bedded, more or less indurated
shales, with abundant pyrite. Large fucoids are common and
animal remains very scarce. At Rockford the same zone is
about 12 feet thick, with little induration, and that confined to
a basal division of about 2 feet, which contains many concretions and much selenite. At Sheffield the presence of the
Striatula zone is questionable, but is indicated by an irregular
belt of gritty, iron-stained, pyritiferous shales lying between the
Sheffield formation and the yellow shales of the Kinderhook.[8]

The Spirifer zone everywhere is characterized by abundant
fossils, and maintains a tolerably uniform thickness of about
20 feet. Within those 20 feet, however, there are marked
differences of lithology and faunas. Unfortunately, the nearly
vertical exposures at Rockford and Hackberry Grove make
accurate measurements of the lower faunules of the zone impossible, while at Bird Hill measurements of the upper ones are

[7] Although we have dropped the name *striatula* for the species of
Schizophoria that is abundant in the Hackberry, we retain the name for
the zone. The reason is that it is quite possible that many paleontologists
will retain their preference for the European name, and use *striatula* instead
of Hall's *iowaensis*.

[8] For a more detailed description of Hackberry stratigraphy, see C. L.
Fenton, *Am. Journ. Sci.*, 4th Ser., vol. 48, pp. 355–376; also in the *American Midland Naturalist*, vol. 6, pp. 179–200. There are, however, certain
points of difference between those papers and the present one; in such
cases the latest statement is the more reliable.

easily available. In the accompanying diagram the relationships of the various minor divisions, or faunules, are indicated. These faunules, as nearly as they can be determined, are described under the heads of five typical localities: the pits of the Mason City Brick and Tile Company and the Northwestern States Cement Company, at Mason City; Hackberry

FIG. 1.   East Portion of Hackberry Grove, the Type Locality of the Hackberry Stage

Grove; the pit of the Rockford Brick and Tile Company, with which is incorporated a hill to the northwestward; and Bird Hill, on the Floyd-Cerro Gordo County line, five miles west and one south of Rockford.

Mason City Brick and Tile Co.:   Feet

2. Atrypa Faunule. Gritty yellow to brown shales, with casts of unidentifiable *devoniana*-like *Atrypas;* also *Schizophoria* and *Spirifer*.   4–6

1. Gracilerectus Faunule.   Gritty, gray-brown, irregularly bedded (commonly heavily bedded) shales with abundant

casts of fucoids. *Gracilerectus hackberryensis* Webster the
typical fossil..

Feet
18–20

Northwestern States Cement Co.:

  2. Soft, shaley, gritty beds of light-gray color. Weather to
clay and chips; fossils almost lacking. May correspond to
No. 2 at Mason City Co. pits; more probably, to part of
No. 1.. 8

  1. Iron-stained, indurated bed, with abundant fucoids....... ½

FIG. 2. The Striatula Zone of the Hackberry at Mason City, showing
the Heavy Bedding and Iron-stained Layers

Hackberry Grove:

  6. Stromatoporella Faunule. Beds in which the stromatoporoids and corals reach their maximum development; *Strophonellas* and broad-hinged *Spirifers* common; pelecypods
at their height. *Floydia* more common than in No. 4.
Thickness about................................... 7

  5. Leptostrophia Faunule. Thin bed of indurated shale with
L. canace and rhizopods abundant. Overlying it are several
inches of laminated shale with few fossils. A wide-spread
and distinct member................................ 1

4. Brachiopod Faunule.[9] Soft shales with the common brachio-   Feet
   pods at the height of their development. Zone of maximum
   abundance of the *Atrypas*.............................   5–7

3. Gigantea Faunule. Yellow shales. Brachiopods abundant;
   characterized by maximum development of gastropods, such
   as *Floydia*, *Bellerophon* and *Straparollus*. Corals less
   common than in Nos. 4 and 6. *Lioclema occidens* rare at
   Hackberry Grove, but abundant a half-mile southeastward.
   With No. 4, totals about 12 feet; thickness.............   5–7

2. Whitneyi Faunule. Gritty, blue beds with casts of *Atrypa*,
   Spirifer, *Schizophoria*, etc.; at the top there are traces of
   No. 3 of the Rockford section.........................   8

1. Indurated, iron-stained beds; concretions common and
   fossils scarce and poor...............................   2

SHEFFIELD FORMATION

Rockford Pit and 'Hill 57':

8. Shales, more or less indurated, weathering to clay and stony
   chips. *Spirifers* of the *orestes* group and *Paracyclas validalinea* the characteristic fossils.......................   1–2

7. Leptostrophia Faunule. At this locality a band of gray-
   buff shaley limestone with *L. canace* abundant...........   ¼

6. Gastropod Faunule. Yellow, calcareous shales with corals
   and gastropods especially abundant. Brachiopods common,
   but less so than in No. 5   5

5. Brachiopod Faunule. Soft, yellow, calcareous shales with
   considerable crystalline gypsum. Brachiopods and rugose
   corals very abundant; zone of maximum abundance of
   Atrypa devoniana. Grades into No. 4, but with thickness of
   about..   6

4. Lioclema Faunule. Shales, blue to yellow, with brachiopods
   abundant, particularly *Spirifer whitneyi* (typical form) and
   S. "hungerfordi." *Lioclema occidens* at maximum abundance,
   as well as size......................................   4

3. Gypidula Faunule. Shales, blue-brown, gritty, with much
   pyrite. Dominant fossils: *Gypidula cornuta* and *Atrypa
   hackberryensis; Douvillina arcuata* common. This horizon
   marks the beginning of the typical Hackberry fauna.....   2–2½

[9] "Hystrix Faunule" of C. L. Fenton, *Am. Mid. Nat.*, vol. 6. p. 185.
The name is changed because *Atrypa hystrix* is not here recognized as a
Hackberry species.

2. Whitneyi Faunule. Shales, blue-brown, gritty, with a few   Feet
brachiopods as casts only. Similar to No. 2 of Mason City
Co. section... 12

1. Concretionary Bed. Indurated gray-brown shales with
abundant calcareous concretions....................... 1–2

In the following section each member has two numbers; the one in parentheses indicates the continuity with the one just previous, the other the sequence at the Bird Hill locality alone:

Bird Hill:

(10) 6. Owen Substage. Residual limestones from the Idiostroma and other zones of the Owen.............. 3

(9) 5. Strophonella Faunule. Shales, soft and yellow, with
fossils very abundant. *Strophonella*, mucronate *Spirifers*
and *Platyrachellas*, and *Reticularia inconsueta* typical.
Stromatoporoids and *Pachyphyllum* at top; rugose
corals abundant throughout...................... 13

(8) 4. *Pugnoides* Faunule. Shale, yellow, hard and lumpy
in bands. Fossils less common than in No. 5 (9);
Strophonella and *Reticularia* relatively scarce; *Pugnoides calvini* the typical species................... 4

(7) 3. Leptostrophia Faunule............................ ¼
(—) 2. Fucoid Faunule. Indurated shale with *Gracilerectus
delicatus* abundant and other fossils scarce.......... 4–5

(6?) 1. Shale, weathering largely to lumps and chips, with
fossils relatively uncommon and poor. Probably corresponds to No. 6 of the Rockford section.......... 3

In this section the correlation has been based on the thin but persistent bed of *Leptostrophia canace* (H. & W.). This same bed is present at Hackberry Grove, thus furnishing a key to correlations with that section. Below it the beds differ very markedly; the heavy, indurated fucoid horizon is quite lacking in the Rockford region. The differences between the upper-most Spirifer zone beds at Bird Hill and those at Hackberry Grove are obvious. These differences, however, are not greater than may be expected in deposits of shallow-water nature, laid down near shore. Furthermore, they are surpassed by the beds of the upper Cedar Valley limestones, where soft gray-blue shale gives way to hard white limestone in a space of three or four hundred yards.

FIG. 3. The Heavy-bedded, Dolomitic Portion of the Owen, at Owen Grove.

Above the dolomite is residuum from the Acervularia zone.

D. RELATIONSHIPS OF THE HACKBERRY FAUNA

Various authors have devoted attention to the age and relationships of the Hackberry fauna. In all cases, however, their efforts have been hampered by inaccurate identifications. Webster and Calvin attempted to refer species after species of the Hackberry fauna to forms already described from the Devonian of New York, but found the operation unsatisfactory. Calvin, too, was much impressed by the similarity of the Hackberry and Independence faunas, saying: "During the time represented by the shales and limestones that lie between the Independence and the Lime Creek [Hackberry] shales the peculiar fauna of the lower shale horizon, adapted to life on a muddy sea-bottom, persisted in some congenial localities at present unknown, suffering in the meantime only a very slight amount of modification, and again appeared, reinforced by a number of new species, when the sea-bottom offered conditiosn

FIG. 4.   The Fucoid Faunule, at Bird Hill.
The shaley beds are crowded with *Gracilerectus delicatus.*

favorable for its success." [10]   Doubtless Dr. Calvin's interpretation is in general correct, but the modification suffered by the fauna is greater than he supposed.   In view of present determinations, it seems that but one or two species are common to the two formations — at least in identical varieties.

A similar resemblance exists between the High Point and the Hackberry faunas, as shown by Clarke.[11]   Again, of course, the number of actual species in common is reduced by later identifi-

[10] *Ia. Geol. Surv.*, vol. 7, pp. 169–170.   1897.
[11] *Bull. U.S. G. S.*, No. 16, pp. 74–76.   1885.

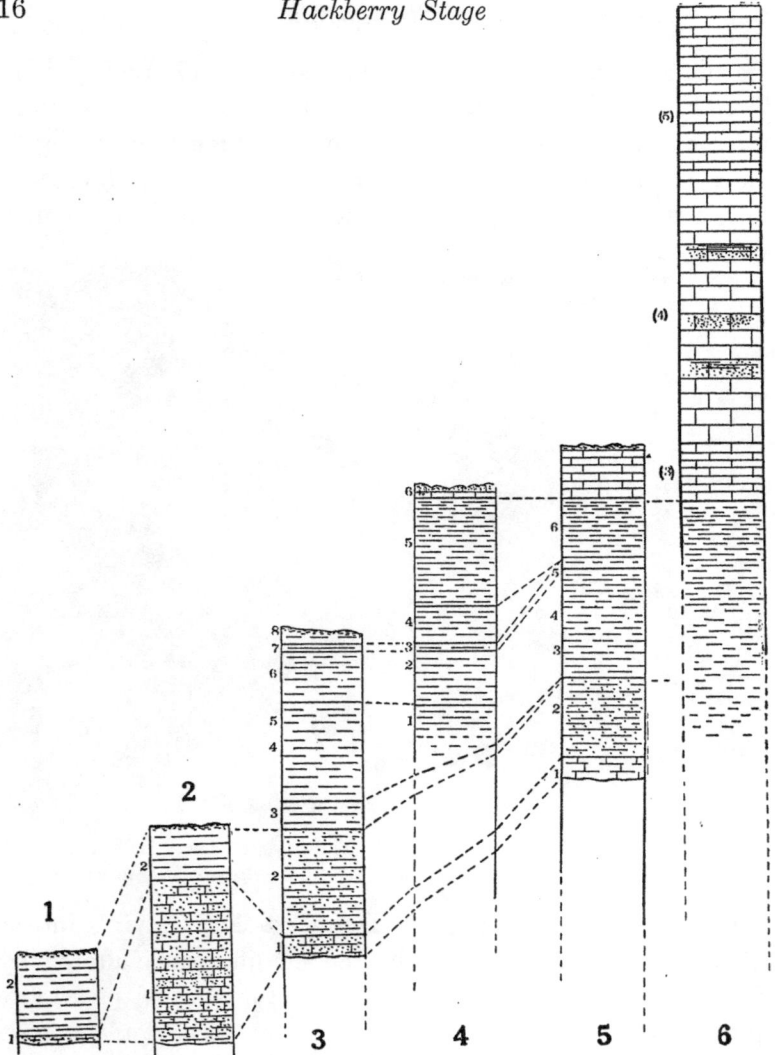

FIG. 5. COLUMNAR SECTIONS OF THE HACKBERRY STAGE

1. Pit of Western States Portland Cement Co., Mason City; 2. "American" Pit, Mason City Brick & Tile Co.; 3. Pit of Rockford Brick & Tile Co., Rockford; 4. Bird Hill Exposure, near Rockford; 5. Hackberry Grove Exposure; 6. Owen Grove, West Exposures.

(Numbers in parentheses correspond to those in the section on page 8.)

cations, but the resemblance is so strong as to indicate that the two faunas had a common origin.

Much the same condition prevails with regard to the Hackberry and other formations. A comparison of the early faunal lists of Webster and Calvin [12] with those of Fenton published in 1919 [13] shows that there has been a progression away from identification of Hackberry species with those of the East and toward the establishing of new species for the Hackberry material. If those 1919 lists are compared with the present paper, a still greater difference becomes apparent. Many Hackberry species *look like* those of the Cedar Valley, the Ouray, the Hamilton, and the Chemung, but close examination shows that, with a few exceptions, they present apparent and constant differences. The same is true of the Cedar Valley faunas, which to a considerable extent have been identified with those of the Hamilton of Michigan and eastern United States and Canada. The lower faunas of the Cedar Valley have a somewhat Hamilton aspect, but the general relationships indicate the formation to be of upper Devonian age, with most of the species of fossils undescribed.

Any conclusions, therefore, regarding the closer relationships of the Hackberry fauna would possess but little value. There are connections with faunas in Nevada, New Mexico, New York, Russia, and northern Canada, but none of them appear to be very close. It will not be until the upper Devonian faunas of the West and Southwest, and particularly the northwestern portions of Canada, are studied and compared with elaborate collections of Hackberry material that reliable opinions may be formed. Until that is done, we may rest the case for the Hackberry fauna with a few general and preliminary working conclusions:

1. The Hackberry stage represents the youngest Devonian known in the Mississippi Valley.

[12] Webster, *Am. Nat.*, vol. 23, pp. 229-243; Calvin, *Ia. Geol. Surv.*, vol. 7, pp. 167-169. 1897.

[13] *Am. Journ. Sci.*, 4th Ser., vol. 48, pp. 368-374; *Am. Mid. Nat.*, vol. 6, pp. 188-197.

2. It possesses a fauna that is at present unique. Such evidence as we possess indicates it to be of western (Asiatic) rather than eastern (European) derivation.

3. That the Hackberry sea appears to have advanced from the northwestward, and to have at one time extended as far eastward as Lake Michigan.[14]

4. That the Hackberry fauna is remarkably prolific, and contains numerous examples of more than usually rapid evolution. This seems to be due to two general causes:

 (a) Life conditions in the Hackberry sea appear to have been extraordinarily favorable for a numerous and varied fauna, in that they
 (i) varied considerably within short distances;
 (ii) varied greatly within short periods of time;
 (iii) afforded ample living space and food for shallow-water organisms.

 (b) Many of the Hackberry species appear to have been in middle or late maturity at the time of their arrival in the Iowa region. These conditions appear to have resulted in
 (i) Extensive production of new species and varieties, such as is normal during the reproductive stage of the life cycle.
 (ii) In some cases — notably among the higher brachiopods — a noticeable production of gerontic forms, especially in late Spirifer zone time.

5. That attempts to correlate the Hackberry with other formations will be more productive of results if the Hackberry fauna is taken as the basis for identification, rather than eastern faunas.

[14] Two species of fish — *Diplodus striatus* and *D. priscus* — are characteristic of the Hackberry, and of the Devonian shales of the Chicago region. Note that Stauffer (*Am. Jour. Sci.*, 5th Ser., vol. 4, pp. 408–411) holds that the Iowa and Minnesota Devonian originated in the Southwest.

II. DESCRIPTION OF FOSSILS

In the following descriptions, the general size of members of any species is expressed in the relative terms large, small, medium, and the like. For these we have used an arbitrary standard based on the assumption that a specimen with a diameter or width of 25 mm. (about 1 inch) is of medium size; one of greater dimension, large; and one of lesser, small.

In the case of new forms, no author name is used after those descriptions to be credited to both C. L. and M. A. Fenton. On the other hand, certain corals described by C. L. Webster and C. L. Fenton are indicated by "Webster and Fenton" following the trivial name.

The location of all type and figured specimens is indicated both in the descriptions and explanations of plates. Casts of many of the Webster specimens, and some others, are in the collections of the following institutions:.

The University of Michigan,
The Walker Museum,
The New York State Museum,
The United States National Museum,
The British Museum (Natural History).

KINGDOM PLANTAE

Phylum THALLOPHYTA

Class ALGAE

" FUCOIDS "

GENUS GRACILERECTUS WEBSTER

Gracilerectus Webster, Am. Mid. Nat., vol. 6, p. 288. 1920.

Description. — "Fossil sea plants or seaweeds, attaining à small to medium size; stems simple, succulent, cylindrical or sometimes compressed; broadly or sharply curved, but sometimes straight; generally distantly branched, branches sometimes opposite; surface smooth or at times irregular; terminations sharp to rounded; root of medium size, flattened or subcircular, generally constricted above, surface smooth or marked by elongated elevations."

Remarks. — In spite of the fact that the name *Gracilerectus* is an adjective, it has been published with adequate definition, and therefore must be used, even though its form renders it highly undesirable. The difficulties and uncertainties of descriptions that seek to apply generic and specific descriptions to such forms as these fucoids are obvious. Yet the fact that they do exhibit distinguishable forms and possess determinable stratigraphic relationships justifies attempts at definition. And even if they are of service only as indices, the genera and species as determined serve a necessary purpose. It is far easier and more satisfactory to refer all Hackberry fucoids to the genus *Gracilerectus* — if we can do no better — than to refer to them merely as fucoids. The same is true of species. Those used and proposed in this manuscript may not be of any great taxonomic value, yet they do designate recognizable types of fossils, and therefore are useful as well as of reliability equal to that of many other species.

GRACILERECTUS HACKBERRYENSIS Webster

Gracilerectus hackberryensis Webster, Am. Mid. Nat., vol. 6, p. 288. 1920.

Description. — "Stem of this seaweed simple, surface nearly even, cylindrical or sometimes compressed, surface smooth so far as known; broadly curved; distantly branched, branches sometimes opposite. Terminations round to pointed. Diameter one fourth to three fourths of an inch; length apparently six inches to two feet or more.

"This fossil, in its usual aspect, presents the appearance of numerous linear stems, often extending half a foot to two feet or more in length, and always appears in the form of casts." [15]

Occurrence. — *G. hackberryensis* is found throughout the lower beds of the Striatula zone at Mason City, lying in great, tangled colonies along the bedding planes.

GRACILERECTUS DELICATUS, n. sp.

(Plate I, Figs. 9–10)

Description. — Stems 7 to 12 mm. in diameter; where suitably preserved they show a single bifurcation 20 or 30 mm. from the base; stems as preserved usually 50 or 60 mm. in length, but appear to represent individual plants of about the same size and proportions as *G. hackberryensis*. Terminations rounded.

[15] Webster, *Am. Mid. Nat.*, vol. 6, p. 288. 1920. The species will be figured in a future number of that journal.

Remarks. — Along with the typical *G. delicatus* at Rockford may be found considerable numbers of fragments of large fucoids, from 20 mm. to 60 mm. in diameter. Inasmuch as there is no structure in either, it is impossible to say whether or not the large and small forms belong to the same species. Since, however, the large forms are lacking in the one bed, the Fucoid faunule, in which *delicatus* is the dominant fossil, such relationship seems impossible.

Occurrence. — *G. delicatus* is characteristic of the Spirifer zone, particularly in the middle and upper beds. At Bird Hill it is the dominant form in the Fucoid faunule, where its stems actually crowd a bed of indurated shale four or more feet in thickness. The branching, however, is seldom well shown, except where small specimens are preserved on flat corals or stromatoporoids.

Holotype. — No. 7775; *Paratypes.* — No. 7776, University of Michigan.

GRACILERECTUS PISTILLUS, n. sp.

(Plate I, Figs. 11–12)

Description. — Stems with a single branching, each branch being club- or pestle-shaped. The total length of the holotype is 64 mm.; that of the allotype, which represents but a single branch, is 90 mm., and the maximum width 38 mm. The club-like form with the greatly enlarged termination, and the short stalk, constitute the distinctive features of this species.

Occurrence. — This species appears to be confined to the portions of the Spirifer zone lying above the Leptostrophia faunule.

Holotype. — No. 7777; *Allotype.* — No. 7778, University of Michigan.

FUCOID (?), gen. and sp. undet.

(Plate I, Fig. 13)

Description. — The form here illustrated is a typical example of a strange 'beaded' fucoid found in the middle portions of the Spirifer zone. It has the general appearance of a cephalopod siphuncle, but the fact that all specimens found are mere shale casts, associated with fucoids, appears to indicate that the fossil represents a plant rather than an animal.

Figured specimen. — No. 26008, Walker Museum.

KINGDOM ANIMALES

Phylum PORIFERA

Class PORIFERA (SPONGIAE)

GENUS CLIONA GRANT

CLIONA HACKBERRYENSIS Thomas

(Plate I, Fig. 8)

Cliona hackberryensis Thomas, Bull. Lab. Nat. Hist. State Univ. of Ia., vol. 6, p. 165, plate. 1911.

Description. — Sponge known only from the burrows, which are tubular and of uniform size, the diameter generally being from .2 mm. to .3 mm. They are most commonly found in *Strophonella hybrida* H. & W., and its varieties, but are not rare in *S. reversa* Hall and *Schizophoria iowaensis* (Hall).

"The ramifying burrows extend parallel to the surface as well as obliquely and vertically to it and are generally filled with some foreign substance [ordinarily clay] which, if softer

than its surrounding walls crumbles out leaving them open;
as the outside of the brachiopod shell weathers away the underlying borings appear on the surface as delicate intersecting
grooves. This labyrinthine maze of passages often weakens the
shell causing it to disintegrate. A single valve containing many
borings was cut and polished, but none of the tubes were found
to perforate the inner surface, showing that in case the sponge
inhabited the shell of a living brachiopod it did not disturb the
occupant in the least." — Thomas, 1911.

Remarks. — We have examined several hundred specimens
of *C. hackberryensis*, and in none of them have we found any
indications of the "pin-shaped siliceous elements" of use in
boring, mentioned in the Zittel-Eastman *Textbook of Paleontology*.
The borings generally contain fine clay, which weathers readily.
Although they are not restricted to the brachial valve, they
are much more common on it than on the pedicle valve of the
host, a fact which seems to indicate that they bored in the
shell during the life of the possessor.

Occurrence. — Throughout the Spirifer zone, being specially
abundant wherever *Strophonella hybrida* is found in great
numbers.

Plesiotype. — No. 7781, University of Michigan.

Phylum COELENTERATA

Class ANTHOZOA

GENUS CHARACTOPHYLLUM SIMPSON

Charactophyllum Simpson, Bull. N. Y. State Mus., vol. 39, no. 8, pp. 209–210, figs. 28–29. 1900.

Description. — Coral irregularly conical or horn-shaped, with complete but thin epitheca. Calyx commonly with diameter less than the maximum diameter of the coral, a condition due to the numerous contractions and expansions of the cylinder during growth. Increase by ova and apparently by calycinal (septal) gemmation. Septa heavy and short, alternating, with the secondaries nearly marginal and the primaries stopping short of the center. The septa are strongly carinate, giving them a denticulate appearance in the calyx which does not, however, show well in the transverse sections. Dissepiments strong; tabulae large and strong, covering the central area, but failing to reach the wall.

Remarks. — The foregoing description has been drawn mainly from the genotype, *C. nanum* (H. & W.), identifiable because it is the only common *Campophyllum*-like coral in the Hackberry. Simpson's description of the genus, too, is worthless, for the carinate septa — the diagnostic character — are not shown in either of his figures. In fact, there is some doubt that the specimen figured by Simpson actually belongs to *C. nanum*, since it lacks the typical irregularity of spacing characteristic of the tabulae of that species. Since, however, accurate identification from the thin section is impossible, a correction, even if necessary, cannot be made.

Occurrence. — Throughout the Spirifer zone, particularly the Strophonella bed, of the Hackberry; according to Simpson, from

the Niagaran of Michigan as well. The reference of *C. radicula*
Rom. to this genus seems questionable, inasmuch as the internal
structure of the type specimens is unknown.

CHARACTOPHYLLUM NANUM (Hall & Whitfield)

(Plate I, Figs. 1–3)

Campophyllum nanum Hall and Whitfield, 23d Ann. Rep. N. Y. State
    Cab. Nat. Hist., p. 232. 1873.
Charactophyllum nanum Simpson, Bull. N. Y. State Mus., vol. 8, no. 39,
    p. 209, perhaps fig. 28. 1900.

Description. — Coral small, elongate subconical or horn-
shaped, irregularly distorted in growth and commonly contracted
in the upper portion. Calyx deep, broad and flattened at base,
with nearly vertical sides; septa strong, alternating, and denticulate, and 60 to 68 in number. Of these, about one-half — the
primaries — reach two-thirds of the distance to the center;
the others are marginal. Tabulae broad and moderately thick;
they are numerous, but very irregularly spaced vertically.
Outer portion with numerous dissepiments.

Remarks. — The illustrations show clearly the characteristics
of this species. When sectioned, it is readily recognizable by
the short septa and the very irregular tabulae; when uncut,
it is distinguished by its irregular mode of growth and general
worn condition of the epitheca, the latter being the more reliable
character.

Occurrence. — Throughout the Spirifer zone; probably rare in
the lower Owen. The commonest rugose coral of the Hackberry.

Plesiotypes. — Nos. 7803 to 7806, University of Michigan.

Genus HELIOPHYLLUM Hall

HELIOPHYLLUM DISPASSUM, n. sp.

(Plate X, Figs. 4–8)

Heliophyllum scrutarium C. L. Fenton, Am. Journ. Sci., 4th ser., vol. 48, p. 370. 1919.

Description. — Coral small, irregularly branching, forming small, ramose coralla consisting of from 2 to 7 corallites. Length of corallites depends upon their position in the colony, and ranges from 5 to 40 mm.; the diameter of the single preserved calyx in the holotype is 10.1 mm. Calyx broad, flattened outwardly, with deep pit. Septa number about 40, are strongly denticulate, and very heavy. They are irregularly alternating; about half of the total number reaches entirely or almost to the center. Tabulae strong, and closely though somewhat irregularly spaced; dissepiments confined to the outer portion of the coral.

Remarks. — This species closely resembles, in general form, *H. scrutarium* Swartz of the Jennings formation of Maryland. Inasmuch as the structure of the latter form is not preserved, accurate identification is impossible. It appears, however, that the Iowa species is larger, more profusely branching, and with a broader calyx than the eastern one. The strong denticulations, branching and heavy epitheca are shown in both. Reproduction is by lateral gemmation.

Occurrence. — Spirifer zone of Hackberry stage.

Holotype. — No. 26045, Walker Museum; *Allotype.* — No. 7809, University of Michigan.

HELIOPHYLLUM SOLIDUM (Hall & Whitfield)

(Plate I, Figs. 4–7)

Zaphrentis solida Hall and Whitfield, 23d Ann. Rep. N. Y. State Cab. Nat. Hist., p. 231, pl. 9, fig. 5. 1873.

Description. — Coral small, subconical to subturbinate, dimensions variable. Calyx deep, broad at bottom, with abruptly ascending sides and weak columella. Septa number 56 to 66; the primaries reach to the center where they coil and form a poorly developed pseudo-columella; secondaries extend only to the central region. All the septa are strongly denticulate, and no specimens show more than a slight trace of the fossula. Tabulae irregular and closely spaced, extending more than half the diameter of the coral; vesicular portion narrow and dense, with the dissepiments small and closely spaced.

Remarks. — The absence of fossula marks this species off from *Zaphrentis*, while the *Cyathophyllum*-like structure and the very strong denticulations show it to be of the genus *Heliophyllum*.

Occurrence. — This typically Hackberry species, which rivals *Charactophyllum nanum* in abundance, is common throughout the Spirifer zone, and the lower portions of the Owen. It is specially abundant in the upper beds of the Spirifer zone at Hackberry Grove and Bird Hill.

Plesiotypes. — Nos. 26002 and 26003, Walker Museum; Nos. 7828 and 7829, University of Michigan.

Genus CHONOPHYLLUM Milne-Edwards & Haime

CHONOPHYLLUM ELLIPTICUM Hall & Whitfield

(Plate III, Figs. 5–8)

Chonophyllum (Ptychophyllum) ellipticum Hall and Whitfield, 23d Ann.
Rep. N. Y. State Cab. Nat. Hist., p. 233, pl. 9, fig. 13. 1873.
Chonophyllum ellipticum Sherzer, Bull. Geol. Soc. Am., vol. 3, p. 169. 1892.
Not *Chonophyllum ellipticum* C. L. Fenton, Am. Journ. Sci., 4th ser.,
vol. 48, pp. 368, 570. 1919.

Description. — Coral of medium size, subturbinate, laterally
compressed and more or less distorted in growth. Calyx moderately to very deep, and of varying configurations; in young
specimens the sides are steep, while in old ones they are nearly
flattened. In the oldest specimen figured the outer portion of
the calyx is much flattened, while the pit is deep, elongate, and
narrow, with vertical sides. Surface, where not weathered,
covered by epitheca which is much wrinkled. Septa number 78
to 102; they show alternation, and are slightly twisted as they
approach the center. There is no columella or pseudo-columella.
Dissepiments are numerous but coarse, and there are broad and
irregular transverse dissepiments crossing the central area. They
are linked up with the septa to such an extent that separation
of them is impossible.

Remarks. — This species is not the one generally identified
as *Chonophyllum ellipticum*, but a much larger, more robust and
heavier form that also is rare. Whether or not Mr. Sherzer's
specimens were of this species is uncertain. All of the specimens
we have examined are in the collection of C. L. Webster.
Among them is a calycle fragment which shows that the diameter of the coral was not less than 65 mm., and that the
calyx was broadly expanded, much like the bell of a horn.

Occurrence. — Middle and upper portions of the Spirifer zone.
Rare.

Plesiotypes. — Collection of C. L. Webster.

Genus TABULOPHYLLUM nov.

Genotype: **TABULOPHYLLUM RECTUM,** n. sp.

Description. — Coral small to large, solitary, irregularly turbinate, subturbinate, or subcylindrical. Growth, in all species, consists of a series of alternating periods of activity and rest. In some of the more symmetrical species these alternations are not pronounced; in others they are so abrupt as to give an appearance of repeated calycinal gemmation. The epitheca is either complete or more or less broken, the latter being the more typical condition. In most species it is thin, and is more or less lacking in weathered specimens. Costae show plainly through the epitheca. Calyx shallow to deep, typically flattened or slightly elevated at the bottom, with sides that ascend at various angles, depending on the species. Fossula very weak or lacking; septa heavy, strong, alternating, non-carinate. Primaries extend entirely or almost to the center; in several species their inner margins unite to form an irregular, vertical tube occupying the central region. In other forms the septa are more or less twisted and coiled, even forming a broad, low pseudo-columella. Commonly there are secondary, irregular calcareous deposits about the septa in the central region.

The tabulae are incomplete, the degree varying with the species. In several forms they are intermingled with dissepiments. Vesicular area commonly well defined; dissepiments small to large, commonly extending into and even across the tabular region. They tend to form broad expansions beyond the main body of the coral. The septa are either free from the tabulae or show upon them, always being less prominent in the zones of crowding, which correspond to the periods of rest.

Remarks. — This genus is, of course, one of the *Cyathophyllidae.* From *Cyathophyllum,* as that genus seems to be considered, it differs in the more imperfect tabulae and inter-mingled dissepiments, the strongly annular type of growth with its attendant bunching of tabulae, the uniformly weak epitheca, and the strong dissepiments which tend to give the corals a vesiculose appearance much like that of *Cystiphyllum*. Indeed, the last named character has proved to be the easiest means of identification of the genus, and by sections has been proved reliable. From *Campophyllum* it differs in manner of growth, strength of dissepiments, and the fact that the septa commonly reach the center and even coil. The septa and tabulae separate it from *Cystiphyllum*.

We are referring to this genus the species called *Cyathophyllum houghtoni* Rom., which Mr. Ehlers considers to be a synonym of *C. traversensis* Winchell. The types show the same characters of tabulae, dissepiments and septa that characterize the Iowa forms. The septa are attached to the tabulae as in *T. erraticum*, and there is a noticeable bunching of the tabulae.

Occurrence. — Spirifer zone and Owen substage in the Hackberry; probably also in the Devonian along the Hay and Athabasca rivers of Canada, and in the Traverse of Michigan.

TABULOPHYLLUM RECTUM, n. sp.

(Plate VI, Figs. 8–12)

Description. — Coral small, irregularly subturbinate, laterally compressed, and distorted. Surface with continuous but thin epitheca which generally is partly eroded, giving that portion of the coral a vesiculose appearance. Transverse sections show that throughout life the growth is compressed; the average proportions show the lesser diameter to be about two-thirds of the greater. Growth somewhat like that of *T. ehlersi*, but with the early creeping less pronounced, and the twisting generally lacking.

Calyx deep, flattened at bottom, with steeply ascending sides. Septa sharp and alternating, 60 to 70 in number, of which about half reach the center. They are thin and weak, and much distorted, and their inner edges unite to form a ring which surrounds the central part of the body. This ring is much distorted and more or less pierced by septa, but no septa reach the exact center, or come into septa from opposite parts of the coral. The tabulae are flattened and nearly complete, and closely spaced, the holotype showing as many as 11 in the space of 2 mm. They exhibit bunching, somewhat as do those of *T. regulare* and the erratic *T. ehlersi*, but the bunches are never widely separated. The vesicular zone is broad and the dissepiments large and coarse; commonly they extend considerably beyond the main body of the coral, forming epitheca-covered frills bearing weak septa. One of these is shown in the transverse section of the holotype.

Remarks. — This species is characterized by the closely and rather regularly spaced tabulae, which extend over about two-thirds of the diameter of the coral, the sharply marked central ring which is without calcareous deposit on its inner side, and the frilling of the dissepiments. In general shape it is about intermediate between *Charactophyllum nanum* and *Tabulophyllum ehlersi*, though weathered specimens may closely resemble *T. regulare* of this paper. It seems probable that the specimen figured in longitudinal section by Simpson as *Charactophyllum nanum* belongs to this species, although it is plain that other specimens considered by Simpson were correctly identified.

Occurrence. — Spirifer zone, and Idiostroma zone of the Owen. Fairly common.

Holotype. — No. 7834; *Allotype.* — No. 7835; Paratypes. — Nos. 7827 and 7836, University of Michigan, and 26046, Walker Museum.

TABULOPHYLLUM REGULARE, n. sp.

(Plate VI, Figs. 1–2)

Description. — Coral small, solitary, subturbinate, and sub-elliptical in transverse section. It expands rapidly above the point of attachment until a diameter of 18 to 22 mm. is reached; above that the size increases but little. Growth consists of a series of progressions and moderate expansions, interspersed with resting periods which are accompanied by constriction of the calyx. At the points of maximum diameter, which coincide with the general duration of the periods of rest, the tabulae are very closely spaced.

Calyx moderately deep, flat-bottomed, with abruptly ascending slopes. Septa number about 70; they are thin and sharp, and alternating. The secondary septa are nearly marginal, while the primaries are produced to the central area where they are considerably distorted and fused at the ends, and do not reach the center. Tabulae irregular and incomplete, and bunched vertically, while the intervening spaces are but slightly filled. Vesicular area wide and rather distinctly separated from the tabular portion; dissepiments large and irregularly spaced, projecting as in *T. rectum*, but to a greater extent. Epitheca originally continuous, but mostly eroded in all specimens examined. In not a few there is none of the epitheca remaining.

Remarks. — This species is closely related to *T. rectum*. It differs from that form in the more pronouncedly bunched tabulae and more open intervening spaces, more sharply distinguished vesicular portion, very marked vesicular extensions, and more regular and less compressed shape. From *T. erraticum* and *T. rotundum* it differs most markedly in the fact that the septa do not encroach upon the tabulae to more than a slight degree. Other differences are noted under those species.

Occurrence. — Spirifer zone, and perhaps in the lower portions of the Owen. Abundant in the Strophonella and Stromatoporella faunules, where it is generally found in short fragments.

Holotype. — No. 7824; *Allotype.* — No. 7825, University of Michigan.

TABULOPHYLLUM EHLERSI, n. sp.

(Plate III, Fig. 4; Plate VI, Figs. 13–16)

Chonophyllum ellipticum C. L. Fenton, Am. Journ. Sci., 4th ser., vol. 48, pp. 368, 370. 1919.
(??) *Campophyllum ellipticum* Whiteaves, Contrib. Can. Pal., vol. 1, pt. 3, p. 202, pl. 27, figs. 5, 6. 1891.

Description. — Coral solitary, small, compressedly subturbinate; invariably much flattened in growth, and distorted, with the attachment prominent. Maximum length about 55 mm.; maximum diameter about 33 mm.; lesser diameter about half that of the greater. Surface covered by a wrinkled epitheca through which the costae show clearly. Calyx shallow and flat-bottomed, with the sides rising abruptly. Septa strong, angular, and alternating; the secondaries are marginal while the primaries reach the center and coil. The total number ranges from 60 to 70. Both primaries and secondaries are weak in the vesicular area. Tabulae very imperfect and irregular; remainder of central region filled by dissepiments, or left empty save for the septa.

From its earliest stages the coral is compressed, and creeps along the surface to which it is attached for as much as 19 mm., with the calyx directed obliquely upward. It then turns upward and expands rapidly but irregularly, showing clearly the influence of periods of rest and growth.

Remarks. — This species, because of its general appearance, has been mistaken for the very rare *Chonophyllum ellipticum,* from which it differs greatly. The specimen sectioned by

Whiteaves appears to be generically related to the Iowa species here called *ehlersi*, but it would appear that Calvin was correct in considering it specifically distinct. We therefore may consider *Campophyllum mcconelli* Whiteaves to be a member of *Tabulophyllum*, closely related to the present species, but different from it in shape and in arrangement and spacing of the tabulae, which are larger in the Canadian form than in the Iowan.

Occurrence. — Throughout the Spirifer zone and in the lower Owen. Whether or not it extends above the Idiostroma zone is uncertain. Most characteristic of the Lioclema faunule at Rockford.

Holotype. — No. 7815; *Allotype.* — No. 7816; *Paratypes.* — Nos. 7817 to 7821, University of Michigan.

TABULOPHYLLUM ROTUNDUM, n. sp.

(Plate II, Figs. 5–10)

Description. — Coral small to medium in size, irregularly subturbinate, solitary, and variously compressed, expanded and twisted in growth. Epitheca continuous, wrinkled and vertically striate; it is partly eroded in most specimens. Calyx roughly circular, flat-bottomed, with steeply ascending sides; depth variable. Septa 60 to 70; the primaries are thin and strong, and reach to the center, where they coil. About the center is a secondary deposit of calcareous material which does not, however, resemble a columella. Secondary septa marginal and weak; they do not show in transverse sections, where even the primaries are indistinct in the vesicular area. Tabulae imperfect, their place being taken to a considerable extent by extended dissepiments; they are never free from the septa, but are more or less so in the areas of vertical crowding or bunching. Vesicular zone occupies more than half the diameter; the dissepiments are large and coarse.

Remarks. — The distinctive characters of this species are several. As contrasted with *T. erraticum, rectum* and *regulare*, it is round rather than compressed in typical expression, expands more rapidly, and is more irregular. From *erraticum* it is distinguished by the septa, which reach the center, by the excessively weak secondary septa, and the coarse vesicular portion. In addition to these ways, it differs from *regulare* and *rectum* in the attachment of the septa to the tabulae, and the imperfect development of the tabulae.

Occurrence. — Spirifer zone of the Hackberry. Uncommon.

Holotype. — Nos. 7838 and 7839, University of Michigan; *Allotype* and *Paratypes.* — Nos. 26004 and 26005, Walker Museum.

TABULOPHYLLUM ERRATICUM, n. sp.

(Plate VI, Figs. 3–7)

Description. — Coral small, solitary, irregularly subturbinate, and compressed and distorted in growth. Calyx moderately deep, flat-bottomed, with rapidly ascending sides; septa number 60 to 70; the secondaries are marginal. Primaries extend to the central region, but stop short of the center where some of them unite. Central cavity bears a calcareous deposit of very irregular form, or may be quite clear. Tabulae imperfect and irregularly distributed vertically; dissepiments large and numerous, projecting as in *T. rectum* and *T. regulare.* Tabulae and central dissepiments generally not free from septa.

Remarks. — This species is characterized by the weak tabulae and the development of septa throughout the central area. The central calcareous deposit appears to be typical.

Occurrence. — Spirifer zone.

Holotype. — No. 26006, Walker Museum; *Allotype.* — No. 7823, and *Paratypes.* — Nos. 7822, 7841 and 7794 to 7796, University of Michigan.

TABULOPHYLLUM EXIGUUM, n. sp.

(Plate III, Fig. 9)

Description. — This species is characterized by a large and very deep calyx, in the center of which is a sharply elevated pseudo-columella. The septa number about 86, are strong, alternating and very widely spaced. The epitheca is heavy and much wrinkled, the form more or less compressed, and the vesiculose appearance marked. The tabulae are very closely spaced — about 3 to the mm. in the portion sectioned; the vesicular zone is narrow, and the dissepiments are directed sharply upward.

Occurrence. — Spirifer zone, upper portions. Rare.

Holotype. — No. 26007, Walker Museum.

TABULOPHYLLUM ROBUSTUM, n. sp.

(Plate V, Figs. 1–3)

Description. — Coral small, subcylindrical, much distorted in growth. Calyx deep, with gradually ascending sides; septa number 74 to 96. The primaries are very strong and heavy, and reach to the center; the secondaries are marginal. The primaries twist sharply in the center, while the center of the calyx is occupied by a deep pit. Epitheca, where present, is strongly wrinkled; over more than half the coral there is no trace of its existence. The attachment is broad and irregular; the dissepiments are large and very closely spaced vertically. Growth expansions, indicative of alternate periods of growth and rest, are numerous — 5 to 6 in the space of 2 cm.

Remarks. — This species is distinguished by the numerous septa, which are very strong, the squat, subcylindrical form, the very coarse dissepiments — many of them are 6 to 8.5 mm. in width — and the deep central pit.

Occurrence. — Middle portions of the Spirifer zone. Rare.
Types. — Collection of C. L. Webster.

TABULOPHYLLUM MAGNUM, n. sp.

(Plate III, Fig. 2; Plate IV, Figs. 1–3)

Description. — Coral of medium size, solitary, subconical or subturbinate; epitheca evidently imperfect or thin, since it is lacking in all specimens. A few specimens are almost sub-cylindrical. The calyx is broad and flat, in mature specimens having a maximum diameter of 60 to 70 mm.; central pit less than half of that diameter, flat-bottomed, and with steep sides. In some specimens there is a slight elevation, but no pseudo-columella. Septa sharp, alternating, and closely spaced; they number 100 to 110 and reach the neighborhood of the center with some twisting but without coiling. This is a distinctive feature of the species, and is well shown in the figure of the holotype.

One of the specimens shows a slight trace of fossula. The tabulae are imperfect, bunched, and interspersed with dissepiments. Tabular and vesicular areas sharply separated in appearance, because of the steep wall of the central pit. Dissepiments fine and closely spaced. Tabulae and central dissepiments connected with septa.

Remarks. — This large species is distinguished by its large, flat calyx, with small, flat-bottomed central pit, and numerous and peculiarly arranged septa. The very closely spaced tabulae and dissepiments, the broad vesicular area, and the sharp divisions into regions distinguish the longitudinal section.

Occurrence. — Owen substage. Generally found in detritus.

Holotype. — No. 26009; *Paratype.* — No. 26010, Walker Museum; two specimens in the collection of C. L. Webster.

TABULOPHYLLUM LONGUM, n. sp.

(Plate III, Fig. 1)

Description. — Coral of medium size, solitary, subturbinate, presumably with thin and irregular epitheca which is lacking in specimens preserved. Calyx 30 to 35 mm. in diameter and half as deep, flat-bottomed, and with steep sides. Septa about 80, alternating, and coiled to form a broad pseudo-columella. The primary septa are much larger and stronger than the secondary. Tabulae incomplete; dissepiments heavy and regularly arranged. Vesicular region narrow.

Remarks. — This species is distinguished by the cone-like shape, very deep calyx, sharply alternating septa and coarse dissepiments. The calyx and width of vesicular area differ markedly from those of *T. magnum*, as well as from *T. expansum*. The form of the coral is so stable, allowing for very ready identification of badly worn fragments, that it clearly is not a variation from some other Hackberry species.

Occurrence. — Owen substage, particularly the Acervularia zone.

Holotype. — No. 26011, Walker Museum, and specimen in the collection of C. L. Webster.

TABULOPHYLLUM EXPANSUM, n. sp.

(Plate II, Figs. 11–12)

Description. — Coral medium to large, broadly subconical, with the upper portion much expanded. Epitheca irregular and lacking in most specimens. Calyx large, irregularly subelliptical; outer (vesicular) portion broad and rounded, with maximum diameter of from 40 to 80 mm. Central pit with a diameter at least half that of the entire calyx; it is shallow and contains a broad pseudo-columellar elevation. Septa 125 to 150, alternating, with the secondaries very weak. The primaries coil, forming a broad and low pseudo-columella. Tabulae small and weak. Dissepiments abundant and closely spaced. In specimens in which some of the epitheca is preserved the costal ridges show distinctly.

Remarks. — This species is characterized by its shortness and its very large calyx, with shallow central pit, close coiling of septa, and very numerous dissepiments. The broad, rounded vesicular area and large pit distinguish the form from *T. magnum*, while the greater number of septa, larger calyx and lesser length, and shallow central pit distinguish it from *T. longum*.

Occurrence. — Owen substage, particularly the Acervularia zone.

Holotype. — No. 26012, Walker Museum.

TABULOPHYLLUM PONDEROSUM, n. sp.

(Plate IV, Figs. 4–5; Plate V, Figs. 5–6)

Description. — Coral large, subturbinate, solitary, covered with an irregularly developed epitheca. The dimensions of the holotype are: length, 117 mm.; width at calyx, 73 mm. The other specimens are somewhat smaller, and less regular. Calyx large and deep, with gently to steeply rising sides; septa 110 to 140, alternating, the primaries reaching to the center and coiling slightly. Tabulae broad, bunched, and more nearly complete than is common in the genus. Vesicular area broad; dissepiments loosely spaced and coarse, projecting more or less as in *T. regulare*.

Remarks. — This is the largest rugose coral known from the Hackberry. It is characterized by the broad, irregularly spaced tabulae, the very coarse dissepiments, and large number of septa. These characters serve to distinguish it from the other large members of the genus, while the size adds one other character separating it from the smaller ones.

Occurrence. — Middle and upper Spirifer zone. Rare.

Types. — Collection of C. L. Webster.

Genus CYSTIPHYLLUM Lonsdale

CYSTIPHYLLUM MUNDULUM Hall & Whitfield

(Plate III, Fig. 3)

Cystiphyllum mundulum Hall and Whitfield, 23d Ann. Rep. N. Y. State Cab. Nat. Hist., p. 234. 1873.

Description. — Coral small, generally less than 3 mm. in length; turbinate, expanding rapidly from the base to the point where it reaches a diameter of about 25 mm., above which the size remains about constant or decreases. Calyx moderately deep, rounded at bottom, with abruptly ascending sides. Septa distinct though thin, and about 60 in number. External surface regular and symmetrical, covered during life by a thin epitheca which seldom is preserved in weathered specimens. Dissepiments closely spaced and heavy.

Remarks. — This species was described, but not illustrated, by Hall and Whitfield. The measurements given by them closely correspond to the specimen in hand, which represents the only species of *Cystiphyllum* which we have found in the Hackberry.

Occurrence. — Spirifer zone, exact horizon not known. The species is rare.

Plesiotype. — No. 26013, Walker Museum.

Genus DIPHYPHYLLUM Lonsdale

DIPHYPHYLLUM TUBIFORME, n. sp.

(Plate II, Figs. 1–4)

Description. — Corallum small to medium in size, formed of irregularly conjoined and more or less distinct corallites. These range in diameter from 4 to 10 mm., and reproduce by lateral gemmation. Each of them has its own complete epitheca. Calyces deep, with steeply ascending sides and moderately thick walls. Septa number 40 to 46; they are in two series, the secondaries reaching into the calyx while the primaries extend almost to the center. Here they either end abruptly, or coalesce to form an imperfect, irregular tube. In the calyx this tube is slightly elevated, giving the appearance of a pseudo-columella. The septa are strongly carinate; tabulae infrequent; dissepiments few and confined to the outer portion of the corallite.

Remarks. — This is a highly variable form in its general shape. Some of the coralla are quite massive, with the corallites in contact, but the larger number are loose masses of corallites of irregular growth. The poorly developed central tube is indicative of *Crepidophyllum*, but it is too poorly developed, and too generally lacking to warrant reference to that genus.

Occurrence. — Spirifer zone, and the Owen substage. Uncommon.

Holotype. — No. 26014; *Allotype.* — No. 26015, Walker Museum; *Paratypes.* — No. 7853, University of Michigan; two specimens in the collection of C. L. Webster.

GENUS STROMBODES SCHWEIGGER

STROMBODES JOHANNI (Hall & Whitfield)

(Plate XV, Figs. 6–7)

Smithia johanni Hall and Whitfield, 23d Ann. Rep. N. Y. State Cab. Nat. Hist., p. 234, pl. 9, Fig. 10. 1873.

Description. — Corallum large, growing in broad, flattened masses. Corallites of moderate size, on the average being distant 10 or 12 mm. from center to center, though the distance varies between 7 and 25 mm. Surface of each corallite slightly concave, with moderate elevation about the central depression. Corallites separated by low but angular elevations. Center ordinarily less than 5 mm. in diameter, shallow, and without well defined columella. Septa rounded and heavy, numbering from 25 to 40, and flexuose. Sections show heavy lamellae interspersed with vesiculose material.

Remarks. — This coral is the largest *Strombodes* of the Hackberry, growing in masses several centimeters in thickness, and 25 or 30 cm. in width. It is characterized by large corallites, which have the borders but slightly elevated about the corallite, and the margin of the central depression little elevated. Epitheca is generally but poorly developed.

Occurrence. — Upper portions of the Spirifer zone and the Owen.

Holotype. — No. $\frac{3720}{1}$ N. Y. S. M.; *Plesiotype.* — Collection of C. L. Webster.

STROMBODES JOHANNI MULTIRADIATUS (Hall & Whitfield)

(Plate XV, Fig. 1)

Smithia multiradiata Hall and Whitfield, 23d Ann. Rep. N. Y. State Cab. Nat. Hist., p. 234. 1873.

Description. — Corallum forming flat expansions; in the holotype corallites are distant 14 to 16 mm. from center to center. Central depression with elevated border, but boundary between corallites not elevated. Septa number 34 to 40, flexuose, and in some cases appearing to join directly with those of other corallites.

Remarks. — This form was described as a separate species, but its appearance hardly warrants the distinction. From *johanni* it differs in larger corallites, slightly coarser, and on the average, more numerous septa, lack of definition in the boundary between corallites, and more noticeable central elevations. None of these features, however, seems to be strongly marked or specially permanent.

Occurrence. — Spirifer zone of the Hackberry, perhaps extending into the Owen. The form is uncommon, and difficult to identify.

Holotype. — No. $\frac{3721}{1}$, New York State Museum.

STROMBODES MARGINATUS, n. sp.

(Plate XV, Figs. 2–5)

Description. — Corallum broad and flattened, forming expansions of undetermined extent and of thickness which varies from 6 mm. to 20 mm., with the average about 15 to 17 mm. Corallites of moderate size, distant 8 to 10 mm. from center to center. Central depression with average diameter of about 4 mm., and depth of 1 mm. or less; columella distinct in weathered specimens. Area about the depression sharply elevated, — in some specimens as much as 2.5 mm. above the general surface of the corallite, which is convex. Septa coarse and rounded; their number varies between 30 and 38, with the prevailing numbers 33 and 34. They are strongly flexuous, and are elevated at the periphery to form distinct boundaries for the corallites. These, however, may be obscured by weathering, or imperfect development in occasional specimens.

Remarks. — This species is distinctly marked by its small corallites, high central elevations, and sharp peripheral ridges. Also, it forms smaller, thinner colonies than does *S. johanni*, and possesses a more fully developed epitheca. In the holotype the epitheca, elevated cells, and point of attachment are well shown, though the allotype has the surface somewhat better preserved. The paratype in the Webster collection differs from the other specimens in a greater regularity of corallite form, coupled with smaller size, and a more regular development of peripheral elevation. The distinction, however, does not appear to be of varietal importance.

Occurrence. — Upper portions of the Spirifer zone and probably in the Owen. It is specially characteristic of the Stromatoporella faunule at Hackberry Grove. It is more common than *S. johanni*.

Holotype. — No. 26053, Walker Museum; *Allotype.* — No. 8086, University of Michigan; *Paratypes.* — No. 8087, University of Michigan, and specimen in collection of C. L. Webster.

GENUS PACHYPHYLLUM MILNE-EDWARDS & HAIME

PACHYPHYLLUM WOODMANI (White)

(Plate VII, Figs. 1–3; Plate VIII, Fig. 2; Plate IX, Figs. 11–12; Plate X, Fig. 3)

Smithia woodmani White, Geol. Iowa, vol. 1, p. 188.   1870.
Pachyphyllum woodmani Hall and Whitfield, 23d Ann. Rep. N. Y. State Cab. Nat. Hist., p. 231, pl. 9, fig. 9.   1873.
Pachyphyllum woodmani Webster, Am. Nat., vol. 23, p. 622.   1889.

Description. — Corallum small to large, growing in irregular, flattened or highly elevated masses, convex circular or oval expansions, or subhemispherical or subovate masses. Small specimens almost invariably show attachment to brachiopods or gastropods; the youngest ones consist merely of two or three corallites attached to such a base. Cell walls slightly to strongly exsert, projecting from .3 mm. to 6 mm. above the intervening spaces; diameter 2.2 to 7 mm., with the average about 5 mm. Entire corallite with diameter of from 3 mm. to 20 mm. Walls thin to thick; periphery generally poorly defined. Central depressions circular to ovate, from 2.5 to 9 mm. in greatest diameter, and .8 to 3 mm. in depth. Septa strong, numbering from 25 to 41, the average being about 34. Half of the septa reach to the center of the depression where there may or may not be an elevation of the pseudo-columella. Tabulae commonly more or less imperfectly developed; dissepiments abundant.

The commonest method of growth of this form is by lateral gemmation, though costal gemmation appears to be not unusual. Epitheca generally lacking, but may be very fully developed, even extending onto the dorsal surface of the colony. Attachment is usually to shells of brachiopods, but corals, bryozoa, and gastropods, as well as stromatoporoids, are among the list of hosts.

Remarks. — This is a highly variable species, as may be seen from the description and illustration. There is great variation in the size of corallites, number of septa and degree of exsertness. In general, the smaller corallites have poorly defined boundaries, exsert surfaces, and septa not exceeding 35 in number. An exception to this is the specimen shown in Plate VIII, Fig. 2, in which the corallites are large and exsert and have as many as 41 septa. The large flat coralla generally have corallites with sharply defined boundaries, slightly exsert borders to the central depressions, and septa numbering from 32 to 40. The corallites, too, are more regular in such specimens.

Occurrence. — This species and its varieties appear to be characteristic of the upper portions of the Devonian of Iowa. One much like the typical Hackberry form, but of varietal difference, appears in the upper Shell Rock beds of the Cedar Valley; in the Nora horizon *Pachyphyllum* is common. It is lacking in the Sheffield and Striatula formations, but appears in the lower Spirifer zone and continues throughout the Hackberry. In the Owen it is more or less replaced by varietal forms.

Plesiotypes. — No. 8090, University of Michigan; Nos. 26016 and 26017, Walker Museum; two polished specimens in the collection of C. L. Webster.

PACHYPHYLLUM WOODMANI ORDINATUM Webster

(Plate VII, Fig. 4)

Pachyphyllum ordinatum Webster, Am. Nat., vol. 23, p. 624. 1889.

Description. — General form and manner of growth similar to that of *P. woodmani.* Corallites regular and well defined, with surfaces concave. Central depression round or ovate, 2 to 4 mm. in greatest diameter; average about 3 mm. Walls heavy and but moderately exsert, projecting 1 to 1½ mm. above the intervening spaces. Septa 28 to 32, half of which extend to the pseudo-columella, while the rest project just into the cup.

In a few cases the exsert portion of the corallite is sunk below the outer margin.

Remarks. — This form differs from the typical *woodmani* in the uniformly smaller central pits, heavy, rounded, and low exsert portion, and convex surface. It may, as Mr. Webster has supposed, represent a distinct species, but to us it seems that a varietal rank is more appropriate, for there is in no respect a clear demarcation from the less exsert forms of *woodmani*.

Occurrence. — Uncommon in the Spirifer zone, upper portions. Collected only at Hackberry Grove.

Holotype. — Collection of C. L. Webster.

PACHYPHYLLUM LEVATUM Webster & Fenton, n. sp.

(Plate V, Fig. 4; Plate XII, Figs. 9–10)

Description. — Corallum massive, irregular or hemispherical. Corallites range from 4 mm. to 12 mm. in diameter, with the greater number measuring approximately 10 mm. Corallites very strongly exsert; the central portions rise as much as 11.6 mm. above the intervening spaces, though most of them do not exceed 7 mm. in height. Septa number 30 to 38; half of them reach to the center of the depression. Upon the elevated portions of the corallites is a heavy ring or band of what appears to be pseudothecalia, which fails to reach the intervening spaces. It is present only on the more exsert corallites, and generally takes the form of a very thick band from 1 mm. to 4 mm. in width.

On the under surfaces of the two specimens examined the point of attachment is very well shown, and there is very little fusion between the corallites. Some of them may be traced from the point of attachment to the periphery, a distance in one case of 66 mm., and in another, of 57 mm. Reproduction is by lateral gemmation, each corallite giving off from 1 to 4 others. Commonly this gemmation is opposite, and the angle made by the young corallite with its parent always is high. Epitheca is developed only in the region of the periphery, where it commonly ascends the marginal corallites. In such cases one may see corallites partly covered by epitheca and bearing, higher up, the pseudothecalial ring.

Remarks. — The validity of this form as a species seems to be established by its constant type. Although but two specimens were examined and measured while the description was being written, several have been collected. All of them show the very high, exsert corallites, the pseudothecalial ring, and the strong individuality of corallites on the ventral side of the corallum. A remarkable feature shown in the holotype is gemmation upon the exsert portions of many of the corallites. Apparently certain of the septa give rise to new corallites, which not only are smaller than the original, but which result in a diminution in the size of the original ones. The secondary corallites bear the pseudothecalial ring, just as do the others. This and the following form are the only members of the genus *Pachyphyllum* so far examined which show lateral, costal, and septal gemmation.

Occurrence. — Middle and upper portions of the Spirifer zone at Rockford and Hackberry Grove. Uncommon.

Types. — Collection of C. L. Webster.

PACHYPHYLLUM IRREGULARE Webster & Fenton, n. sp.

(Plate X, Figs. 1–2; Plate XI)

Description. — Corallum massive, irregularly subhemispherical, with maximum diameters 12 to 14 cm. Corallites irregular, with diameters of 13 to 20 mm. for the larger ones, and 6 to 8 mm. for the ones arising by costal gemmation. Inner wall strongly exsert, rising as much as 11 mm. above the intervening spaces; they are sharply inclined away from the center of the coralla, giving the whole a radial appearance. Central depression generally at one side of the corallite; round or oval, with diameter more than half that of the corallite. Depression about two-thirds the diameter in depth, except in the smaller corallites, where it is about one-third. Septa strong, 37 to 46 in number, with the average about 40; corallite boundaries poorly defined. Epitheca generally lacking.

Remarks. — This form is related to *P. levatum*. It shows the same septal gemmation, although less commonly than does that species. Many of the corallites bear poorly developed pseudothecalial rings, and nearly all of them show constriction. Even those not showing the ring have decided breaks in the septa. The large size of the corallites, the exceedingly coarse appearance, and the different form of corallum, along with the outward direction of the corallites, serve to distinguish the species.

Occurrence. — Middle and upper portions of the Spirifer zone. Rare.

Types. — Collection of C. L. Webster.

PACHYPHYLLUM OWENENSE Webster & Fenton, n. sp.
(Plate VII, Fig. 5)

Description. — Corallum flattened, discoid; never so large nor so thick as the more massive coralla of *P. woodmani*. Corallites 12 to 24 mm. in diameter, with the average about 17 mm. Central portion but moderately exsert, with a shallow depression, the diameter of which ranges from 7 to 9 mm. Septa strong, numbering 36 to 42, with the average 40; lines of demarcation between corallites poor, though the septa are not confluent.

Remarks. — This species, probably descended from the same line that produced *P. ordinatum*, and perhaps rising from the large-celled forms of *P. woodmani* like the one shown in Pl. VII, Fig. 2, is typical of the Acervularia zone of the Owen, where it almost replaces *P. woodmani.* The flattened coralla, the very large, low-walled corallites with their shallow central pits, and the large number of septa seem to be stable characters which do not intergrade with the typical *P. woodmani.* The nearest relative appears to be *P. crassicostatum* Webster, from which it differs in having smaller and less exsert corallites, which show a lesser tendency to be distinct, a more compact *woodmani*-like manner of growth, and much flatter and more regular coralla.

Occurrence. — Owen substage, particularly the Acervularia zone. Common, but generally poorly preserved.

Holotype. — No. 26054, Walker Museum; *Allotype.* — Collection of C. L. Webster.

PACHYPHYLLUM CRASSICOSTATUM Webster

(Plate VIII, Fig. 1; Plate IX, Figs. 1–5)

Pachyphyllum crassicostatum Webster, Am. Nat., vol. 23, p. 623. 1889.

Description. — Corallum simple, irregularly branching, or massive, depending upon age, and from 2 to 11 cm. in diameter. Corallites from 7 to 22 mm. in diameter, with central depressions occupying from one-half to four-fifths of the diameter. The former is the case in corallites belonging to massive coralla; the latter in those belonging to branching ones. Diameter of corallites increases with the age of the corallum.

The septa number 31 to 60, though 38 to 42 are the commonest numbers in mature coralla. Half of the septa stop just within the exsert inner wall, the rest extending to the perpendicularly perforate pseudo-columella. In large specimens the

central depression contains a pit 1 to 2.3 mm. in depth instead
of an elevation; in smaller ones this pit is less distinct. Ventral
surface of corallum without epitheca, the septa continuing unbroken from the periphery to the point of attachment. In
young coralla the corallites are quite distinct, while in the large
coralla the outer corallites are more or less united.

Remarks. — This species is distinguished by the large corallites, irregular manner of growth, and the characteristic pseudo-
columellar pit. In type of growth it is the most variable of the
Iowa members of the genus *Pachyphyllum.*

Occurrence. — Owen substage, particularly the Acervularia
zone. Not uncommon in the dolomitic beds of the middle Owen
as well.

Cotypes. — Collection of C. L. Webster; *Plesiotypes.* — Nos.
26019 and 26020, Walker Museum.

PACHYPHYLLUM CRASSICOSTATUM NANUM, n. var.

(Plate IX, Fig. 6)

Pachyphyllum crassicostatum, variety, Webster, Am. Nat., vol. 23, p. 623.
1889.

Description. — Corallum forming a small, irregularly branching mass; corallites 9 to 14 mm. in diameter, the average being
about 12 mm. Exsert portion occupies about four-fifths of the
surface of the corallite; diameter of the central depression
about two-thirds that of the corallite. The depression contains
a slight elevation, there being no trace of a pit such as is found
in the typical *crassicostatum.* Septa number 35 to 40, 37 being
the commonest number. As in other species of *Pachyphyllum,*
half of them reach to the center, while the rest stop within the
inner wall. Epitheca consists of a few irregular, broken bands.

Remarks. — The adult coralla of this variety somewhat resemble a young *crassicostatum,* although the corallites are more

fully united than in that species. The corallites never attain the great size of those of *crassicostatum*, nor do the coralla become large. In general type, this variety might well be ancestral to the Owen form.

Occurrence. — Middle and upper beds of Spirifer zone at Hackberry Grove and Rockford. Webster refers to specimens from the upper Spirifer beds at Owen Grove. Uncommon.

Holotype. — No. 26018, Walker Museum.

GENUS MACGEEA WEBSTER

Macgeea Webster, Am. Nat., vol. 23, p. 710. 1889.

Description. — Coral simple, cylindrical or horn-shaped, commonly more or less flattened and somewhat distorted by attachment and irregular growth. Length ranges from 2 mm. in very young specimens to 50 mm. in adults of very old age; diameter 1.5 mm. to 20 mm.; calyx subcircular in young and subovate or distorted in old specimens.

Calyx variable in proportions, but usually about equal in depth and width. Outer wall thin; epitheca heavy but not continuous, and never reaching the border of the calyx. Septa number 32 to 70, with sharp alternation in size. Costae continuous with septa.

Remarks. — This genus is more or less related to *Pachyphyllum*, and the genotype was placed in *Pachyphyllum* by the original authors. The solitary manner of growth, septa which fail to approach the center in adult specimens, and very strongly developed tabulae serve to distinguish the genus.

Occurrence. — So far, members of this genus are known only in the Independence Shales and the Hackberry stage of the Iowa Devonian.

MACGEEA SOLITARIA (Hall & Whitfield)

(Plate IX, Figs. 7–10)

Pachyphyllum solitarium Hall and Whitfield, 23d Ann. Rep. N. Y. State
Cab. Nat. Hist., p. 232, pl. 9, figs. 6–7. 1873.
Macgeea solitaria Webster, Am. Nat., vol. 23, p. 711. 1889.

Description. — Coral solitary, cup-like or horn-shaped; length
2 to 50 mm., and maximum diameter 1.5 to 20 mm. Lower
extremity generally preserves attachment; in many specimens
the attachment distorts the entire coral. In the young specimens the transverse section is nearly circular, but with increased
size there comes decided flattening, so that it is not uncommon
to find specimens in which the maximum diameter of the calyx
is twice that of the minimum.

Calyx generally as deep as wide, but not always so in young
specimens. Outer wall thin; septa 55 to 70, alternating in size.
In the youngest stages, or those portions of the coral representing the youngest stages, the primary septa reach entirely or
almost to the center, and show a slight tendency to form a
pseudo-columella. In specimens upwards of 6 mm. in length
the septa stop short of the center, and in those 12 to 15 mm.
in length there is no trace of columella. Secondary septa very
short, reaching just within the calyx, or extending forward and
uniting with the primaries. Costae continuous with the septa,
extending 2 to 10 mm. below the edge of the calyx. Rest of
the surface bears an irregularly developed but heavy epitheca,
which may be either banded or continuous.

Remarks. — *Macgeea* is a common genus in the Hackberry,
being most commonly represented by the genotype, *M. solitaria.*
Sections show certain substantial differences from the illustrations of Hall and Whitfield; there is never a strong pseudo-
columella, while the tabulae are much stronger than those
indicated in the original figure. Nor, as we have pointed out,

do the septa reach the center in any but young stages, while dissepiments are more numerous than Hall and Whitfield's Figure 7 indicates. Apparently their longitudinal section, Figure 6, was made from a specimen not ground to the center, for we found that in grinding our sections a stage was reached when they represented the published section. The transverse section probably is diagrammatic, from a poor specimen.

Occurrence. — Throughout the Spirifer zone, but probably not in the Owen. Colonial and common, particularly in the Strophonella faunule at Bird Hill.

Plesiotypes. — Nos. 7798, 7800, and 7801, University of Michigan.

MACGEEA CULMULA Webster

Macgeea culmula Webster, Am. Nat., vol. 23, p. 712. 1889.

Remarks. — This name was proposed for a form smaller and more elongate than *M. solitaria*, and with 30 to 32 septa. The dimensions of the holotype are stated to be: diameter, 4 mm.; length, 22 mm. We have not seen the types nor found specimens that agreed with the definition. Inasmuch as there is no illustration with the description, we are inclined to question the validity of the description.

GENUS ACERVULARIA SCHWEIGGER

In the Hackberry are several forms referable to the problematic genus *Acervularia*. Somewhat related species have been referred by other authors to *Cyathophyllum* and *Prismatophyllum*, on the grounds that *Acervularia* is not a good genus. In the preparation of this volume there has not been time to undertake any full examination of sections, yet from such data as we possess there seems no valid reason for abandoning the genus.

The following species, therefore, are placed in *Acervularia* until there is opportunity for a detailed study of the relationships of that supposed group and its relatives.

ACERVULARIA INEQUALIS Hall & Whitfield

(Plate XIV, Figs. 7–8)

Acervularia inequalis Hall and Whitfield, 23d Ann. Rep. N. Y. State Cab. Nat. Hist., p. 233, pl. 9, figs. 11–12. 1873.
Not *Cyathophyllum inequale* Swartz, Md. Geol. Surv., Lr. Dev., p. 205, pl. 20, figs. 1–4. 1913.

Description. — Corallum massive, convex to subhemispherical, or forming flattened expansions with thicknesses up to 4 cm. Corallites closely united, polygonal, increasing by interstitial gemmation. Their diameters vary from about 3 mm. in the smallest to 7 mm. in the largest ones, the average being in the neighborhood of 4.5 or 5 mm. Central depression with a diameter about half that of the corallite, and a depth about half its diameter. Septa slightly corrugated on sides and edges. In the larger corallites they number 27 to 30; in the smaller, 24 to 26. The commonest number is 28. Walls between the cells strong and sharply elevated. About half the septa reach the center, the rest stopping just within the central depression. Tabulae rather weak but abundant; dissepiments closely spaced.

Remarks. — This species differs sharply from the one generally called *A. inequalis*, and here named *whitfieldi*, in having much smaller and more regular corallites, fewer septa, and much flatter coralla. The original description is sufficiently clear, so that there is no doubt that it was the smaller species which furnished the type for *inequalis*.

Recently Swartz, in the reference given, made this species synonymous with *Columnaria* (*Cyathophyllum* or *Prismatophyllum*) *inequale* of Hall, from the Helderberg of New York, New

Jersey and Maryland. As may be seen by comparing the figures of the Iowa form with those of Swartz's specimens, there is but little similarity between the two, and certainly no reason for placing them in the same species.

Occurrence. — Upper portion of the Spirifer zone, particularly in the Stromatoporella faunule at Hackberry Grove.

Plesiotypes. — No. 26048, Walker Museum, and a specimen in the collection of C. L. Webster.

ACERVULARIA WHITFIELDI, n. sp.

(Plate XIV, Figs. 1–3)

Description. — Corallum massive, convex to subhemispherical, forming masses whose diameters vary between 2 and 25 cm., depending mainly upon age. Corallites of varying sizes, the smallest (which are uncommon) having diameters of 3 to 4 mm., and the largest, 13 mm.; the average is between 6 and 7 mm. Central depression occupies about half the diameter and is less than half as deep as wide, with the center slightly elevated. Septa strong, corrugated on sides and edges. They number 29 to 32 in the smallest corallites, and 32 to 36 in the larger ones. The commonest number appears to be 33. About half the septa reach into the central depression, while the remainder extend to the margin or slightly beyond. Walls well marked, forming sharp ridges which separate the corallites. Dissepiments closely spaced vertically.

Remarks. — This species generally is identified as *A. inequalis,* because it is the one that most collectors have secured in quantity. It differs from that species in the greater size of the cells, great variation in their size within a single corallum, and in the notably larger number of septa. The manner of growth, too, is much more massive than that of *S. inequalis,* and the coralla are much larger.

Occurrence. — Spirifer zone, especially the Stromatoporella faunule. Uncommon or rare in the lower portions of the Owen; rare in the Acervularia zone.

Holotype. — No. 5319, University of Michigan; *Allotype* and *Paratype.* — Nos. 26055 and 26040, Walker Museum.

ACERVULARIA BASSLERI Webster & Fenton, n. sp.
(Plate XIII, Fig. 2; Plate XIV, Figs. 4–5)

Description. — Corallum massive, growing in broadly sub-ovate, subhemispherical, or loaf-shaped masses, which attain diameters of 20 cm. or more, and thicknesses upward of 12 cm. The large allotype figured on Plate XIII has a maximum diameter of 16.6 mm. and a thickness of about 4.8 mm., while the holotype measures 21.8 cm. × 17.6 cm. × 8.6 cm., and is not an unusually large specimen.

Corallites irregularly hexagonal or septagonal, with maximum diameters ranging from 6 to 16 mm., the average maximum being about 12 mm. Septa 38 to 43, strong and rugose. External walls somewhat zigzag, and elevated from .2 mm. to 1.2 mm. above the general surface of the corallites. Surfaces of corallites strongly funnel-shaped, after the manner of *A. profunda* of the Cedar Valley, but to a lesser degree. Central depression subelliptical, its diameter generally being less than half that of the entire corallite. The secondary septa extend slightly beyond the border of the depression, while the primary ones extend to the center, where they form a weak pseudo-columella. In a few cases the coralla possess corallites that have grown to considerable length, their surfaces being covered with a heavy epitheca. One such specimen is illustrated. In such corallites the individuals are quite distinct; each calyx is separated from the others, and surrounded by its own epitheca, which generally is heavy.

Polished sections show the walls between the cells to be
strong, while the septa possess a somewhat wavy appearance.
They are heavy near the walls, and thin rapidly toward the
centers, where the primary ones form the pseudo-columella.
Dissepiments are strong but not abundant; tabulae distantly
spaced.

Remarks. — This moderately variable species belongs to the
general group of *Acervularia profunda* Hall of the Cedar Valley.
From that species it differs in having larger and more uniform
corallites, finer and smoother septa, more regular walls, and
calyces of a different type. In *profunda* the depression of the
calyx begins at the boundary of the corallite, and continues
without pronounced break. There is no definite central depression; the whole calyx is depressed. Also, the columella is
less distinct in the Cedar Valley species than in the one from
the Owen. It appears, therefore, that all references to *A.
profunda* in the Hackberry are in error.

There may be some question as to the reliability of some of
the species of corals here proposed, particularly on the part of
those who are disposed to give a free rein in the interpretation
of the word 'species.' We must admit that corals are highly
variable; that often two forms which do not look alike prove
to interbreed. Unfortunately, however, we lack such criteria
in paleontology, so that it seems justifiable to describe as species,
varieties, etc., any types that are sufficiently distinct to admit
of ready identification. And such is the case with the 'species'
of *Pachyphyllum*, and the species and varieties of *Acervularia* of
this volume.

Occurrence. — *Acervularia bassleri* appears in the Idiostroma
zone and continues throughout the Owen, becoming specially
common in the Acervularia zone.

Holotype, Allotype, and two *Paratypes.* — Collection of C. L.
Webster; *Paratype.* — No. 8085, University of Michigan.

ACERVULARIA BASSLERI DEPRESSA Webster & Fenton, n. var.

(Plate XII, Figs. 1–2)

Description. — Coralla much like those of *A. bassleri,* but never attaining so great size. Corallites with diameters of from 5 to 11 mm., the average varying with different specimens. A few specimens, which appear to belong to this species and variety, show diameters as great as 14 mm. for some of the corallites. The number of septa ranges from 33 to 42, the commonest numbers approximating 42.

The distinguishing things about the variety are the small and very deep corallites, which commonly have depths equal to or more than half their widths. Like the corallites of *A. profunda,* these have the central depression occupying most of the surface, although there generally is an outer, flattened, elevated portion. The pseudo-columella is weaker than in the typical *bassleri,* and the septa are more regular and much more rugose.

Remarks. — The variety *depressa* appears to be a local development from *A. bassleri.* In extent of differentiation it varies, some coralla appearing much like the parent species, while others are very distinct. In a few cases, coralla that clearly belong to *bassleri* bear a few corallites that closely resemble those of *depressa.*

Occurrence. — Upper portions of the Owen, particularly the *Acervularia* zone, where it is common though usually poorly preserved.

Holotype and *Paratype.* — Collection of C. L. Webster; *Allotype* — No. 8084, University of Michigan.

ACERVULARIA BASSLERI MAGNA Webster & Fenton, n. var.

(Plate XIV, Fig. 6)

Description. — The general characters of this form are shown clearly in the figure. Most of the corallites are very large — 22 to 25 mm. in greatest diameter — and very irregularly polygonal. The septa are coarse and distantly spaced, numbering 44 to 52 in the large corallites. In the small ones, which occur near the periphery, the number is 40 to 45. The outer portion of the corallite is flat, while the border between the corallites is high, heavy, and crenulate. The diameter of the central pit is one third to one half that of the entire corallite; the pit is abrupt, and occupied at the bottom by a broad, low, pseudocolumella, which is composed of somewhat less than half of the septa.

Remarks. — This form is marked by its large corallites, numerous, coarse septa, and the flattened outer area. It seems quite probable that it deserves rank as a separate species, but the number of specimens is so small that no reliable estimate of the permanence of the distinctive characters has been possible.

Occurrence. — Acervularia zone of the Owen; uncommon.

Holotype. — Collection of C. L. Webster.

Genus ALVEOLITES Lamarck

ALVEOLITES ROCKFORDENSIS Hall & Whitfield

(Plate XII, Figs. 3–6; Plate XIII, Fig. 1)

Alveolites rockfordensis Hall and Whitfield, 23d Ann. Rep. N. Y. State Cab. Nat. Hist., p. 229. 1873.

Description. — Corallum forming broad, irregular, discoidal or oval expansions that range in size up to a maximum of more than a foot on the longest diameter. The thickness is most commonly less than an inch, but in some specimens is as much as three. Corallites small, numbering from 9 to 12 in the space of 5 mm. Septa very slightly developed; in most specimens indistinguishable. Apertures rhombic, highly oblique, with the middle of the upper lip forming a sharp, subangular elevation, and that of the lower occupying the angle between two corallites in advance.

Remarks. — This species is one of the commonest and most persistent corals of the Hackberry. Although not figured by Hall and Whitfield, there can be little doubt as to the species. The authors' statement, however, that there are 40 to 50 cells "in the space of one-tenth of an inch" appears to be an error in writing or printing, for about that number of corallites are to be found in an inch.

Alveolites rockfordensis is a gregarious species, but does not form reefs. It grew among stromatoporoids, and many specimens show interlaminations of *Stromatoporella* and *Syringostroma* along with the coral. In such specimens, the *Alveolites* furnishes the base, the stromatoporoids being of later growth. One such specimen, in the portion of our collections which is located at the Walker Museum, shows numerous layers, and a total thickness of 3 inches (77 mm.).

Occurrence. — Throughout the Hackberry from the lower part of the Spirifer zone to the top of the Owen limestones. It is particularly common in the middle beds at Rockford and the *Stromatoporella* faunule at Hackberry Grove; also in the *Acervularia* zone of the Owen at various localities.

Plesiotypes. — Nos. 7844 to 7846, University of Michigan; No. 26056, Walker Museum.

GENUS CLADOPORA HALL

CLADOPORA FLOYDENSIS, n. sp.

(Plate XII, Figs. 7–8)

Description. — Corallum ramose or (rarely) palmate; branches generally 5 to 7 mm. in diameter, with frequent bifurcations that form angles of 70 to 90 degrees. Corallites nearly round in some specimens, but more generally irregularly polygonal; about 7 in the space of 10 mm. Walls stout, thickening near the periphery. Orifices oblique, but with walls that do not project more than a fraction of a millimeter above the general surface. In weathered specimens the corallites appear quite distinct, and the projecting lips are lacking. Lateral pores very sparsely distributed; diaphragm occasionally indicated but rarely preserved.

Remarks. — This species typically occurs as broken fragments of ramose coralla, but a few specimens show that the general form was, in isolated examples at least, palmate. The corallites are large in comparison with the size of the branches, and the walls but little elevated. Internal structure generally lacking.

Occurrence. — Basal Spirifer zone through the Owen, becoming less common in the upper portions of the formation.

Holotype. — No. 7849; *Allotype.* — No. 7850; *Paratype.* — No. 7851, University of Michigan.

Genus AULOPORA Goldfuss

AULOPORA SAXIVADUM Hall & Whitfield

(Plate XVI, Fig. 8)

Aulopora saxivadum Hall and Whitfield, 23d Ann. Rep. N. Y. State Cab. Nat. Hist., p. 235, pl. 10, fig. 6. 1873.

Description. — Corallum prostrate, branching, attached to the surface of some other organism, most commonly a coral or brachiopod. Bifurcation commonly at every other corallite, though at each corallite, and at the third, it is not infrequent. Angle of bifurcation ranges from 30 to about 120 degrees, 90 degrees appearing to be the average. Tubes range from .7 mm. to .9 mm. in diameter, slender at base and enlarging rapidly. Apertures round, elevated .2 to .5 mm. above the prostrate portion of the tube and from 1 mm. to 1.5 mm. above the base of attachment; wall about the aperture slightly thickened; tubes wrinkled throughout their length.

Remarks. — This species is associated with, and closely resembles, the preceding one. It differs from *A. iowaensis* in the smaller size of tubules, lesser height of aperture, infrequent coalescing of tubules, and thinness of the tubules at the apertures. It does not exhibit the large colonial growth of *iowaensis*, nor is it commonly found on the shells of *Floydia*. *A. saxivadum* is most abundant on rugose corals, as well as on brachiopods, particularly *Schizophoria* and *Strophonella*.

Occurrence. — Common throughout the the Spirifer zone of the Hackberry; less common in the Owen,

Plesiotype. — No. 7783, University of Michigan.

AULOPORA ADNASCENS, n. sp.

(Plate XVI, Figs. 3–4)

Description. — Corallum branching, prostrate, attached to surfaces of bryozoans, apparently being restricted to *Lioclema occidens* (H. & W.). Tubules branch at every second or third corallite; in section they are round with diameters of .5 to 1 mm. They are narrow at the bases and increase gradually toward the apertures. Average corallites 2.5 to 3 mm. in length; apertures .5 mm. to 1 mm. above base of attachment and directed obliquely or upward.

Remarks. — This species differs from *A. paucitubulata* in that the corallum branches more frequently and the apertures are considerably larger.

A. adnascens is remarkable because of its close association with *Lioclema*, and commonly is intergrown with it, with only the apertures protruding to the surface. The surface of one of these specimens is illustrated.

Occurrence. — Throughout the Spirifer zone, particularly at Rockford, where *L. occidens* is most abundant.

Holotype. — No. 7789; *Paratype.* — No. 7790, University of Michigan.

AULOPORA IOWAENSIS Hall & Whitfield

(Plate XVI, Fig. 2)

Aulopora iowensis Hall and Whitfield, 23d Ann. Rep. N. Y. State Cab. Nat. Hist., p. 235, pl. 10, fig. 5. 1873.

Description. — Corallum branching, prostrate or forming heaped up elevations. The tubes commonly coalesce and overgrow one another, so as to obscure the surface to which they are attached. Cells 2 to 3 mm. in length and .9 to 1.4 mm. in width, with large round apertures that are elevated from 1 to 3 mm., and are directed upward so that the aperture is almost parallel to the plane of the attached portion of the cell. In all cases there is a distinct enlargement and thickening of the wall as it nears the aperture, so that in an aperture measuring .5 mm. the diameter of the tube is 1.2 mm. Bifurcations frequent, usually one for every corallite. Tubes wrinkled by growth marks throughout their length.

Remarks. — This species is found particularly on shells of the large gastropods here referred to the genus *Floydia*, although it has been found on corals, stromatoporoids, and brachiopods as well. The average specimens show the vertical portions of the tubes considerably worn, but occasional ones, such as the plesio-type, show portions which were protected from wear prior to fossilization.

Occurrence. — Found throughout the Spirifer zone, from its base to the Strophonella faunule. It is most common in the Floydia faunule at both Hackberry Grove and Rockford, where it is found on almost every gastropod whose shell is preserved. Probably to be found in the Owen substage, but identifications are open to some question.

Plesiotype. — No. 26047, Walker Museum.

AULOPORA MAXIMA, n. sp.

(Plate XVI, Fig. 7)

Description. — Corallum prostrate, branching, attached to the surface of other organisms, particularly rugose corals. The holotype shows branching at every third cell. Length of cells from 6.5 to 8 mm.; diameter of the tubes 1.6 to 3 mm.; slender at base and enlarging rapidly toward the aperture. Apertures round, from 2.8 to 3.9 mm. in diameter; directed upward or obliquely; slightly elevated above the prostrate portion of the tubes and from 2 to 3 mm. above the base of attachment. Tubes annulated throughout their length.

Remarks. — This is the largest *Aulopora* in the Hackberry as well as the rarest.

Occurrence. — Spirifer zone; rare.

Holotype. — 7788, University of Michigan.

AULOPORA EXPANSA, n. sp.

(Plate XVI, Fig. 10)

Aulopora annectens C. L. Fenton, Am. Journ. Sci., 4th ser., vol. 48, p. 369. 1919.

Description. — Corallum prostrate, branching, attached to the surface of some other organism, most commonly a stromatoporoid. Bifurcations not numerous; angles of bifurcation vary from 15 to 90 degrees or more. Tubes round or slightly flattened, from 1.9 mm. to 2.1 mm. in diameter; wide at base and enlarged very gradually up to the region of the aperture, where they broaden abruptly. Aperture large, round or oval, ranging from 1.7 to 2 mm. in diameter; elevated 1 to 1.7 mm. above the base of attachment. Shell thickened around the base, cup flattened in the bottom and walls thick.

Remarks. — This species closely resembles *Aulopora annectens* Clarke (*Bull. 16, U. S. G. S.*), to which it originally was referred. It differs from that form, however, in the total absence of pseudo-septa, as well as in the rounder tubes.

Occurrence. — Lower Owen.

Holotype. — No. 26021, Walker Museum; *Paratype.* — No. 7786, University of Michigan.

AULOPORA INCRUSTANS, n. sp.

(Plate XVI, Fig. 1)

Description. — Corallum prostrate, attached to surface of a stromatoporoid and completely covering both sides of the specimen. Tubes average 2.5 mm. in length and 1.5 mm. in diameter. One, two or three corallites are given off around the aperture of the mother cell; the tubes increase in size to the aperture. Apertures round, thick-walled, and directed obliquely upward; about .5 mm. in diameter. Tubes usually wrinkled and coalesced so that the entire surface of the stromatoporoid is hidden.

Remarks. — This species resembles *A. iowaensis* in that the tubes coalesce but in manner of growth it is very different. In *A. incrustans* the distinctly flattened tubes creep along the surface of another organism and direct their apertures usually obliquely, while the apertures of *A. iowaensis* are directed upward from 1 to 3 mm. above the prostrate portion of the corallite.

Occurrence. — Holotype, only specimen in our collections, is from the Spirifer zone at Hackberry Grove.

Holotype. — No. 8091, University of Michigan.

AULOPORA PAUCITUBULATA, n. sp.

(Plate XVI, Fig. 9)

Aulopora saxivadum of some collections identified by C. L. Fenton.

Description. — Corallum branching, prostrate, most commonly attached to the upper surfaces of *Syringostroma* and other flat-growing stromatoporoids. Tubules branch every third or fourth cell; diameter .8 mm. to 1 mm. Cells 2 mm. to 3 mm. in length; apertures circular, slightly elevated, and commonly directed obliquely rather than directly upward. Tubules expand little or not at all anteriorly, and are most commonly constricted near the aperture; coalescence infrequent. Entire tubules coarsely wrinkled.

Remarks. — This species differs from the two preceding in size and type of apertures. The characters are well shown in the figures.

Occurrence. — The upper portion of the Spirifer zone, particularly at Hackberry Grove (Stromatoporella faunule), where it is to be found on specimens of *Syringostroma*.

Holotype. — No. 7791, University of Michigan; *Allotype.* — No. 26029, Walker Museum.

AULOPORA MINIMA, n. sp.

(Plate XVI, Figs. 5–6)

Description. — Corallum prostrate, attached to other organisms, particularly rugose corals. Holotype shows only one branching, which is dichotomous. Cells 1.5 to 2 mm. in length; tubes .6 mm. in diameter, slightly constricted at bases and usually swollen at the apertures; smooth except for few wrinkles. Apertures round, smaller than the tubes, directed obliquely, and slightly elevated above the prostrate portion.

Remarks. — This species differs from *A. paucitubulata* in that it is much smaller and averages 8 apertures to the centimeter instead of but 5 as in the latter.

Occurrence. — Spirifer zone; rare.

Holotype. — No. 7788, University of Michigan.

GENUS SYRINGOPORA GOLDFUSS

SYRINGOPORA ROCKFORDENSIS, n. sp.

(Plate XVI, Fig. 11)

Description. — All specimens of this species are broken masses of tubes, apparently representing coralla which originally had a branching form of growth typical of the genus *Syringopora*. The tubules are irregular, branching, and with an average diameter of about 1.5 mm.; the connecting tubules range from .6 to 1 mm. in outside diameter. Distance between the tubules uncertain, but apparently less than 2 mm. on the average. Tubules vermiform, expanding rapidly; one with a length of about 11.5 mm. has a diameter at the lower end of about .9 mm., and at the upper, of 1.8 mm. Radial crests indistinguishable; diaphragms directly transverse where they are preserved.

Remarks. — This is the common *Syringopora* of the Hackberry. It is found in broken masses in the harder portions of the shales. The preservation usually is poor, the internal characteristics being almost indistinguishable.

Occurrence. — Throughout the Spirifer zone, particularly in the region of Rockford, where the fossils are preserved somewhat better than at the northwestern localities.

Holotype. — No. 7785, University of Michigan; *Paratypes.* — No. 26047, Walker Museum.

Phylum MOLLUSCOIDEA

Class BRYOZOA

Genus ASCODICTYON Nicholson & Etheridge

ASCODICTYON PUSTULUM, n. sp.

(Plate XVII, Fig. 2)

Description. — Zoarium attached to surfaces of brachiopods, particularly *Spirifer* and *Schizophoria.* Vesicles isolated or clustered without definite arrangement; semi-hemispherical or semi-ovate; of various sizes with maximum diameters of .7 mm. or even more. Delicate stolons radiate from isolated vesicles or connect groups of vesicles; in specimens less well preserved they are wanting. Surface of vesicles marked by pores.

Remarks. — *A. pustulum* differs from *A. stellatum* Nicholson and Etheridge in that it has no definite arrangement of vesicles and seems to have fewer pores in the vesicles.

Occurrence. — Spirifer zone.

Holotype. — No. 8035, University of Michigan; *Allotype.* — No. 26032, Walker Museum.

Genus VINELLA Ulrich

VINELLA (?) PANDA, n. sp.

(Plate XVII, Fig. 1)

Description. — Zoarium attached to brachiopod shells; consists of irregularly ramifying tubular stolons arranged in an irregularly radial manner, or ramifying without apparent pattern. Stolons average .3 mm. in diameter, enlarging somewhat, and extend 1 to 4 mm. without branching or uniting with other stolons. Surface of type too badly weathered to show pattern, or pores if they are present.

Remarks. — This problematical fossil is referred to the genus *Vinella* Ulrich. It lacks the typically radial form of that genus, and the presence of pores on the tubes is uncertain.

Occurrence. — Spirifer zone; not common.

Holotype. — No. 8038, University of Michigan.

Genus HEDERELLA Hall

HEDERELLA ALTERNATA (Hall & Whitfield)

(Plate XVII, Figs. 5–6)

Stomatopora (?) *alternata* Hall and Whitfield, 23d Ann. Rep. N. Y. State Cab. Nat. Hist., p. 235, pl. 10, figs. 7–8. 1873.

Description. — Zoarium branched and attached to surface of other organisms; consisting of a tubular axis from which the zoœcia are given off alternately to the left and right, though occasionally two are given off in succession on the same side. Zoœcia tubular, round or oval, indistinctly annulated, diameter .3 to .5 mm. Seven zoœcia given off from the main axis in the space of 5 mm. Apertures terminal, equal to the width of the zoœcia, usually oval and opening upward but not elevated.

Occurrence. — Throughout the Spirifer zone and in the lower portion of the Owen substage. It is found on corals, brachiopods and molluscs, but particularly on brachiopods of the genus Spirifer.

Plesiotype. — No. 8039, University of Michigan.

Genus HERNODIA Hall

HERNODIA HALYSON, n. sp.

(Plate XVII, Fig. 4; Plate XVIII, Fig. 4)

Description. — Zoaria irregularly branched, prostrate, attached to surface of other organisms, particularly brachiopods. Zoœcia oval; bud from preceding zoœcia at angles of about 45 degrees; are usually the same size throughout though there may be a slight constriction or swelling at the bases. Zoœcia .5 to .7 mm. in diameter; 2 mm. or less in length; 5 to 7 in 5 mm. Apertures terminal, round or oval, slightly smaller than the cells; directed obliquely or vertically, most commonly the former, and elevated only slightly above the prostrate portion of the zoaria.

Occurrence. — Middle and upper portions of the Spirifer zone, at Rockford, Bird Hill, and, probably Hackberry Grove.

Holotype. — No. 8041; *Paratypes.* — Nos. 8042 and 8043, University of Michigan, and 26035, Walker Museum.

HERNODIA LINEARIS, n. sp.

(Plate XVII, Fig. 7)

Description. — Zoaria irregularly branched, prostrate, attached to other organisms, particularly *Floydia concentrica.* Zoœcia oval; bud from preceding zoœcia at angles of from 10 to 15 degrees; are the same size throughout or gradually become larger. Zoœcia .4 mm. in diameter at the terminus; .8 to 1.4 mm. in length; 8 to 10 in the space of 5 mm. Apertures oval, the same size as the zoœcia, directed obliquely, but not raised above the prostrate portion of the zoaria.

Remarks. — *H. linearis* differs from *H. halyson* in smaller diameter and greater number of zoœcia. The most distinctive character, however, is the lesser degree at which one zoœcium buds from the preceding. The apertures of *H. linearis* are directed obliquely and never raised above the prostrate portion of the zoaria; the surfaces of the zoœcia are smooth while those of *H. halyson* are wrinkled.

Occurrence. — Spirifer zone.

Holotype. — No. 26033, Walker Museum; *Allotype.* — No. 8040, University of Michigan.

Genus PETALOTRYPA Ulrich

PETALOTRYPA FORMOSA, n. sp.

(Plate XVII, Figs. 8–10; Plate XVIII, Figs. 1–3)

Description. — Zoarium of bilaminar, narrow, undulated branches, sometimes appearing to inosculate. Measurements of holotype and a paratype: height, 15 mm. and 14 mm.; width, 4.8 mm. and 12.8 mm.; thickness, 1.9 mm. and 2 mm. Basal expansions are found on crinoid stems, brachiopods and Fenestellas; the cells are of the same general type as those of the bifoliate branches. The paratype has monticules from 2.5 to 2.9 mm. apart; they consist of cells of average or larger than average size, with thicker walls. Cell apertures polygonal, commonly quite regularly quadrangular; about 9 in the space of 2 mm. Mesopores subangular, not numerous and tending to gather in clusters. Zoœcial tubes straight in the mature region; prostrate portion of tubes short in the immature. Round or subangular median tubulae are found in the mesotheca. Diaphragms few or wanting. Acanthopores small and rather scarce.

Remarks. — This species differs from *P. delicata* Ulr. in that it has monticules, thinner zoœcial walls and a different shape. In growth much like *P. compressa*, but the latter species has larger zoœcia and a more robust zoarium.

Occurrence. — Upper portions of the Spirifer zone, particularly the Strophonella faunule.

Holotype. — No. 8050; *Paratypes.* — Nos. 8051 to 8054, University of Michigan.

GENUS LIOCLEMA ULRICH

LIOCLEMA OCCIDENS (Hall & Whitfield)

(Plate XVII, Fig. 11; Plate XVIII, Figs. 5–6)

Fistulipora occidens Hall and Whitfield, 23d Ann. Rep. N. Y. State Cab.
　Nat. Hist., p. 228, pl. 10, figs. 9–10. 1873.
Callopora cincinnatiensis Ulrich, Journ. Cin. Soc. Nat. Hist., vol. 1, p. 93,
　pl. 4, fig. 8–8a. 1878; ibid, vol. 5, p. 142, pl. 6, figs. 18–18a. 1882.
Leioclema occidens Ulrich, Geol. Surv. Ill., vol. 8, p. 426. 1890.

Description. — Zoarium of irregularly branching masses, variable in form, from 6 to 8 cm. or even more in length, 5 to 7 cm. in width. Surface smooth or bearing monticules from 2 to 4 mm. apart. Zoœcia round or oval, thin-walled, 6 to 7 in the space of 2 mm.; encircled by a series of round or subangular mesopores that are open at the surface. These are about one-half to two-thirds as large as the zoœcia. Diaphragms not numerous in the zoœcia and very few in the mesopores; they are irregularly placed in either case. Acanthopores small, numerous on well-preserved specimens.

Remarks. — *L. occidens* is the most common Hackberry bryozoan, and is gregarious. Many specimens have *Aulopora adnascens* imbedded in the substance so that only the apertures of the coral are showing.

Occurrence. — Throughout the Spirifer zone; numerous at most localities, but very rare at Hackberry Grove.

Plesiotypes. — Nos. 8061 and 8062, University of Michigan.

LIOCLEMA MINUTUM (Rominger)

(Plate XVIII, Fig. 7)

Fistulipora minuta Rominger, Proc. Acad. Nat. Sci. Philadelphia, p. 120. 1866.
Leioclema minutum Ulrich. Geol. Surv. Ill., vol. 8, p. 427. 1890.

Description. — This species closely resembles *L. occidens* (H. & W.) in surface characters, but is distinguished from it by zoarial habit. *L. minutum* (Rom.) occurs as thin incrusting layers .2 to .3 mm. in thickness; found on Fenestellas and brachiopods. Zoœcia oval or subangular and 5 to 7 in the space of 2 mm.; mesopores subangular, open at the surface and entirely encircling the zoœcia. Acanthopores quite numerous and seem to encroach upon the zoœcia.

Remarks. — *L. minutum* is one of the fossils, and the only bryozoan to be identified with forms outside of the Hackberry. Inasmuch as Dr. Rominger possessed a collection of Hackberry fossils, the authors suspect that the type of *L. minutum* accidentally got among Michigan material from the Iowa sets.

Occurrence. — Spirifer zone; rare.

Plesiotype. — No. 8069, University of Michigan.

Genus FENESTELLA Lonsdale

FENESTELLA DIATRETA, n. sp.

(Plate XVIII, Figs. 15–16)

Description. — Zoarium flabellate; largest fragment seen 8 mm. in length and 10 mm. in width. Obverse usually with straight, slender branches that bifurcate at distant intervals; branches .3 mm. in width and 11 to 12 in the space of 5 mm. Dissepiments short; about the same width as the branches.

Fenestrules elliptical or subquadrangular, from .3 to .5 mm. in length and about .2 mm. in width; 6 in the space of 5 mm. Carinae poorly developed, bearing small irregularities, but not nodes. Zoœcia in two ranges. Apertures small, circular, usually 6 to each fenestrule; those between the fenestrules larger than those on the sides. On the reverse the branches and dissepiments are rounded and of the same character as those of the obverse, and the dissepiments are on the level with the branches. Fenestrules are elliptical or subquadrangular and form straight rows. Surface of both branches and dissepiments is smooth and marked by very fine, irregularly placed pits.

Remarks. — This species has been commonly identified as *F. vera* Ulr., but is distinguished from the latter by the poorly developed carinae, absence of distinct nodes and the straight branches on the reverse, marked by fine pits.

Occurrence. — Throughout the Spirifer zone.

Holotype. — No. 8071; *Allotype.* — No. 8072; *Paratypes.* — Nos. 8073 and 8074, University of Michigan.

FENESTELLA VERA DISSIMILIS, n. var.

(Plate XVIII, Fig. 13)

Description. — Zoarium flabellate, slightly undulating; holotype measures 12 mm. in length and 10 mm. in width. Obverse with straight branches that bifurcate at distant intervals. Branches .2 mm. or a little more in width and 11 to 13 in 5 mm. Dissepiments short, about two-thirds the width of the branches and a little below them; they have distinct ridges, but they are by no means as prominent as the carinae of the branches. Fenestrules subquadrangular, measuring about .6 by .3 mm. and 6 or 7 in 5 mm. Carinae prominent, quite sharp, bearing small nodes about their diameter apart; on a well-preserved portion of surface 5 were found in the space of 1 mm. Zoœcia in 2 ranges. Apertures small, circular and opening obliquely into the fenestrules; 6 to 8 to each fenestrule and 22 to 23 in 5 mm.

Remarks. — This variety differs from *F. vera* Ulr. in that it has smaller branches, slightly larger fenestrules, probably sharper carinae, and smaller nodes.

Occurrence. — Spirifer zone.

Holotype. — No. 8076, University of Michigan.

FENESTELLA CARINATA, n. sp.

(Plate XVIII, Fig. 14)

Description. — Zoarium flabellate, slightly undulating; measurements of holotype and allotype: length, 18 mm. and 9 mm.; width, 13 mm. and 8 mm. Obverse with straight branches that bifurcate at intervals of 3.5 to 5 mm. Branches average .4 mm. in width, .5 mm. in thickness and 11 in the space of 5 mm. Dissepiments short, considerably less than the width of the branches and 5 in the space of 5 mm.; usually below the surface of the branches. Fenestrules elongate rectangular, averaging .2 mm. in width and 1 mm. in length, 4 in the space of 5 mm. Carinae elevated, slightly zigzag but do not bear distinct nodes. Zoœcia in two ranges, apertures circular, 8 to 10 to a fenestrule and 19 to 20 in the space of 5 mm. On the reverse the branches correspond to those of the obverse and together with the dissepiments form quite regularly subrectangular fenestrules. Dissepiments the same size as the branches and on a level with them.

Remarks. — This species is closely allied to *F. diatreta*, but is distinguished from it by the longer fenestrules, wider branches and zigzag character of the carinae.

Occurrence. — Spirifer zone.

Holotype. — No. 8077; *Allotype.* — No. 8078, University of Michigan.

GENUS ORTHOPORA HALL

ORTHOPORA BUCHERI, n. sp.

(Plate XVIII, Figs. 10–12)

Description. — Zoaria very slenderly branched, from .4 to 1 mm. in diameter; bifurcations few. Cells tubular, arising at the center of the branch and diverging gradually until near the surface, when they turn outward rather abruptly. Apertures oval, 6 in 2 mm., arranged in rows longitudinally and alternating so that they are in diagonal rows as well; 14 to 16 rows on one branch. The rows of apertures are separated by well defined ridges on which are acanthopores. No mesopores.

Occurrence. — Spirifer zone.

Holotype. — No. 8079; *Allotype.* — No. 8080, University of Michigan.

ORTHOPORA ULRICHI, n. sp.

(Plate XVIII, Figs. 8–9)

Description. — Zoarium ramose, branches very slender, from 1.2 to 2 mm. in diameter. Zoœcial apertures oval to subangular; from 7 to 9 in the space of 2 mm. Zoœcial tubes approach the surface by a steady curve from the median portion of the branch; diaphragms irregularly placed in the immature regions. Cell walls increase in thickness from the main axis to within a short distance from the surface, from which point they decrease. Thus in weathered portions the walls appear to be thick and in fresh surfaces very thin. Mesopores few, acanthopores numerous.

Occurrence. — Spirifer zone.

Holotype. — No. 8082, University of Michigan.

Class BRACHIOPODA

Genus CRANIA Retzius

CRANIA FAMELICA Hall & Whitfield

(Plate XIX, Figs. 1–3)

Crania famelica Hall and Whitfield, 23d Ann. Rep. N. Y. State Cab. Nat. Hist., p. 236, pl. 11, figs. 6–7. 1873.

Description. — Shell small, flattened to subconical. Diameter varies from .8 mm. and less in young specimens to 17 or 18 mm. in adults. The largest and most perfect specimen figured has a total thickness, from plane of base to apex, of 6.5 mm.

Shell thin, the brachial valve generally more or less crushed or flattened in young, and lacking in mature specimens, or so badly crushed as to show little of its shape. In the young it almost invariably is marked by rib-like elevations, which more or less correspond to the plications of the host. Thus when the host is *Atrypa devoniana* the ribs are coarse, while on *Spirifer hungerfordi* they are considerably finer. Surface also marked by concentric, lamellose lines of growth, which are specially prominent on the pedicle valve. This amounts to little more than an incrustation upon the surface of the host.

Remarks. — *Crania famelica*, as might be suspected from its habits, is a highly variable form, and perhaps is deserving of separation into at least two varieties. Those shells which are to be found on Spirifers always are smaller than the ones on corals, stromatoporoids, and Atrypa; the ones on *Strophonella hybrida* and its varieties are the largest of all. Unfortunately, the largest specimens fail to show much of the character of the brachial valve. Because of the necessary uncertainty it has seemed undesirable to make any division of Hall and Whitfield's species.

The following is a list of the organisms on which *Crania famelica* is most commonly found:

Tabulophyllum ehlersi	Strophonella reversa (and vars.)*
Heliophyllum solidum	Strophonella hybrida (and vars.)*
Pachyphyllum woodmani	Atrypa devoniana *
Charactophyllum nanum	Spirifer hungerfordi *
Syringostroma planulatum	Spirifer whitneyi (and vars.)*
Crania famelica	Floydia gigantea *
Schizophoria iowaensis	Floydia concentrica *

Those species marked by an asterisk are the most common hosts to this endoparasitic brachiopod.

Occurrence. — Rare in the Gypidula faunule, common in the Spirifer zone, and rare in the Owen.

Plesiotypes. — Nos. 7904 to 7906, University of Michigan.

CRANIA LIOSOMA, n. sp.

(Plate XIX, Fig. 4)

Description. — Shell small, discoid, broadly subconical, with the apex sharply elevated. Diameter of the holotype about 9.5 mm.; thickness 3.2 mm.

From the species *C. famelica* this form differs in greater thickness of shell and complete lack of rib-like markings. Concentric lamellae very pronounced. On the holotype, which retains much of its coloration, the central portion is brown, with faint olive mottling, while on the outside is a band, averaging 2 mm. in width, of dark olive-brown color. This band is separated from the rest of the shell by heavy growth lamellae.

Remarks. — This species is quite distinct from *C. famelica.* All of the free specimens approximate in size the one figured; they lack ribs, and in nearly every case show traces of coloring.

The lamellose lines are specially prominent, and serve to define the outer band of darker color.

Occurrence. — Middle and upper portions of the Spirifer zone. An uncommon species.

Holotype. — No. 7907, University of Michigan.

CRANIA STEWARTI, n. sp.

(Plate XVII, Fig. 3)

Crania crenistriata Hall and Clarke, Pal. N. Y., vol. 8, pt. 1, pl. 4H, fig. 12 only. 1892.

Description. — Shell small, flattened, or depressed-subconical; greatest diameter of the holotype 10.7 mm.; lesser diameter 8.4 mm.; apex sharp and subcentral. Surface of brachial valve marked by fine radiating striae, which are weak or lacking near the apex and increase in size and number toward the periphery, where they number about 4 or 5 to the mm. Concentric growth lamellae heavy, and irregularly spaced.

Remarks. — The holotype of this species was referred, with doubt, to Hall's *Crania crenistriata*, and so illustrated by Hall and Clarke. From that species it differs in having much finer striae, heavier apex, and less regular form. The near absence of striae in the central portion is another distinctive feature. The two forms are, of course, closely related, but as suggested by Dr. Clarke, are not identical.

Occurrence. — Very rare; known only from the middle portions of the Spirifer zone.

Holotype. — No. 11894, Walker Museum.

GENUS SCHIZOPHORIA KING

SCHIZOPHORIA IOWAENSIS (Hall)

(Plate XIX, Figs. 5–11)

Orthis iowensis Hall, Geol. Ia., vol. 1, pt. 2, p. 488, pl. 2, fig. 4. 1858.
Orthis (Schizophoria) impressa Calvin, Ia. Geol. Surv., vol. 7, p. 167. 1897.

Description. — Shell large; wider than long with the greatest width near the mid-length of the shell; hinge-line equal to one-half to two-thirds the greatest width; cardinal extremities rounded. Dimensions of three specimens: length of pedicle valve, 13.6 mm., 20.9 mm. and 26 mm.; width, 16.9 mm., 25.9 mm. and 33.9 mm.; thickness, 8.6 mm., 13.5 mm. and 18 mm.; length of hinge-line, 9.4 mm., 14.4 mm. and 20 mm.

Pedicle valve moderately convex in umbonal region, curving rather abruptly to cardinal margin and gently to the antero-lateral and anterior margins; flattened near the cardinal extremities and on the lateral slopes. Mesial portion is slightly flattened anteriorly in young specimens; in the mature ones it is depressed to form a wide shallow sinus beginning about 10 mm. anterior to the beak. Beak pointed, incurved; cardinal area of moderate size, the curvature increasing near the beak, with the flatter portions making an angle of 90 degrees to the plane of the valve. Delthyrium triangular. The dental teeth are supported by short dental plates that continue as low, rounded or subangular ridges around the muscular scars; muscular scars subovate in outline, one-half or less than one-half the length of the valve; muscular area divided longitudinally along the median line by a rounded ridge, which is highest at its termination and is gradually reduced in height toward the beak. Not infrequently the median ridge is represented by a slight thickening of the shell anterior to the muscular scars. Margin of the shell marked by costae.

Brachial valve highly convex; greatest convexity posterior to the middle; the surface curves very abruptly to the cardinal margin and but slightly less so to the lateral and anterior margins; compressed toward the cardinal extremities. Mesial portion not distinguished from the general curvature, with the shell broadly and indefinitely flattened from the umbonal region anteriorly. The umbo extends a little beyond the cardinal margin; the beak small and incurved; the cardinal area small. Cardinal process small and angular; crural plates prominent, widely divergent, sockets deep. Muscular scars subcordate in outline, bounded laterally by irregular angular ridges and anteriorly by low but distinct regular semicircles; divided longitudinally by a subangular ridge that originates about 4 mm. from the base of the cardinal process.

Surface of both valves marked by fine radiating costae, about 4 in the space of 2 mm., that increase by bifurcation and implantation. These costae are tubular, and weathered specimens frequently show ovate punctae at irregular intervals. Concentric growth lines fine and usually numerous.

Remarks. — This species is one of the group generally referred to *S. striatula* of Schlotheim. That species, as illustrated by the specimens we have examined, differs from the Iowa form in more and finer plications, thinner shell, less prominent muscle scars, and much stronger vascular markings in the brachial valve. Because of these differences we are abandoning the name *striatula* for the particular species under consideration, and reviving Hall's well defined and illustrated species name *iowaensis*.

Occurrence. — In the Hackberry, *S. iowaensis* appears as casts in the lowermost Striatula faunule, becomes very abundant in the Spirifer zone, and disappears in the Owen. A form closely related, if not identical, is rather common in the Nora limestone, which in this region is the uppermost formation of the Cedar Valley stage. The exact relationships of a number of *Schizo-*

phorias from other upper Cedar Valley horizons have not been determined.

Plesiotypes. — Nos. 7908 to 7914, University of Michigan.

SCHIZOPHORIA IOWAENSIS, form MAGNA nov.

(Plate XIX, Figs. 12–18)

Description. — Shell has the same general shape as *S. iowaensis*, but is much heavier. Dimensions of three typical specimens: length of pedicle valve, 22 mm., 26 mm. and 30 mm.; width, 16.1 mm., 32.3 mm. and 38.5 mm.; thickness, 19 mm., 22.3 mm. and 27.5 mm.; length of hinge-line, 14.7 mm., 18.7 mm. and 18.8 mm.

Pedicle valve has its margin rounded by growth lamellae, as does the brachial valve, which makes the juncture of the two very indistinct. The hinge-line is shorter in proportion to the size of the shell and the cardinal extremities are more rounded than those of *S. iowaensis*. Lingular extension of the mesial sinus directed upward more acutely than in *S. iowaensis*. Pedicle valve heavy, with a distinct thickening around the interior margin. The rostral portion of the shell shows the same characters as in *S. iowaensis*. The muscular scars not infrequently show growth lamellae toward the anterior. The median ridge dividing the muscular area is sometimes represented anterior to the muscular scars by a slight thickening of the shell. Vascular areas marked by elongate nodes and ridges. Lateral and anterior margins marked by fine costae.

Brachial valve very strongly convex in the middle. Umbonal region elevated higher above the cardinal area than that of *S. iowaensis*. Surface curves abruptly to cardinal margins, very steeply to lateral and anterior margins. No conspicuous compression in the postero-lateral margins. Cardinal extremities very rounded. Slight flattening of the mesial portion from the umbo anteriorly. Cardinal area small, beak incurved.

Surface of both valves marked by the same type of costae as those in *Schizophoria iowaensis*. Anterior region shows the tubes in the costae particularly well.

Remarks. — *Schizophoria iowaensis magna* seems to reach old age when much smaller than *S. iowaensis* and has a heavier shell throughout life. It is distinguished by its great convexity and heavy shell.

Occurrence. — Same as *S. iowaensis.*

Holotype. — No. 7920; *Paratypes.* — Nos. 7921 to 7926, University of Michigan.

Genus STROPHEODONTA Hall

STROPHEODONTA THOMASI, n. sp.

(Plate XX, Figs. 1–5)

Description. — Shell small, concavo-convex, with greatest width along or slightly anterior to the hinge-line. Dimensions of three specimens, the third of which is the holotype: length of pedicle valve, 3.4 mm., 4.7 mm. and 9.5 mm.; length of brachial valve, 3.2 mm., 7.3 mm. and 8.3 mm.; width along hinge-line, 3.6 mm., 8.6 mm. and 10.6 mm.; thickness, 1 mm., 4.5 mm. and 4.5 mm.

Pedicle valve highly convex, greatest convexity near the mid-length. Beak elevated. A slight constriction at the cardinal extremities forms auriculations on most specimens. Cardinal area of the holotype 1 mm. in height; broadly triangular, horizontal or slightly arched and vertically striate; denticulations do not reach the cardinal extremities. Deltidium about as high as wide, smooth and indistinct. Teeth quite prominent. Muscular area bounded laterally by 2 heavy ridges that parallel the margins of the shell and become indistinct anteriorly. Diductor scars large and flabelliform; adductor scars subtriangular, the posterior element being transversely ovate, distinctly bounded laterally, and moderately deep, while the anterior is triangular and indistinct. Median septum low, broad and extending to about the mid-length of the shell. Lateral and anterior areas pustulose; vascular markings on the anterior areas.

Brachial valve concave; cardino-lateral areas slightly flattened. Cardinal area narrow and smooth. Cardinal process bifid, prominent and curved backward. Dental plates small and the sockets shallow. The posterior adductor scars are small, indistinct depressions anterior to the base of the cardinal process. The anterior adductors are very small and are bounded laterally by the short, low, rounded crural plates. The median septum originates in front of the anterior adductors and anteriorly becomes narrow, rather high and rounded. Pustules mark the surface on either side of and anterior to the muscular scars.

Surface of both valves marked by coarse, rounded or subangular costae; from 13 to 18 about 2 mm. from the beak. Their number increases by bifurcation and implantation.

Occurrence. — Throughout the Spirifer zone, but particularly in the middle portions. Rare in the Gypidula faunule.

Holotype. — No. 8028; *Allotype.* — No. 8029; *Paratypes.* — Nos. 8030 to 8032, University of Michigan; 26079 and 26080, Walker Museum.

DOUVILLINA ARCUATA (Hall)

(Plate XX, Figs. 6–10)

Strophodonta arcuata Hall, Geol. Ia., vol. 1, pt. 2, p. 492, pl. 3, figs. 1a–1c, 2a, b, e, f. 1858.

Strophodonta arcuata Calvin, Bull. U.S. Geol. and Geog. Surv., vol. 4, p. 728, 1878.

(?) *Strophodonta arcuata* Whiteaves, Contrib. Can. Pal., vol. 1, p. 285. 1892.

(??) *Strophodonta arcuata* Walcott, Mon. U.S. G. S., vol. 8, p. 121. 1884.
Stropheodonta arcuata Hall and Clarke, Pal. N. Y., vol. 8, pt. 1, p. 289, pl. 15 B, figs. 1 to 3. 1892.
Stropheodonta arcuata Schuchert, Bull. U.S. G. S., No. 87, p. 419. 1897.
Douvillina arcuata, C. L. Fenton, Am. Mid. Nat., vol. 5, pp. 213, 215. 1918.

Description. — Shell small to below medium size; semicircular to semielliptical in outline. The ellipticity is more pronounced in adult specimens. Hinge-line equal to or greater than the width at the mid-length of the shell. Dimensions of a very perfect small specimen, a medium-sized one and a large one with the cardinal extremities broken off: length of pedicle valve, 8.8 mm., 14 mm. and 16.9 mm.; width, 9.6 mm., 17.8 mm. and 23.3 mm.; thickness, 6 mm., 8 mm. and 8 mm.; height of cardinal area, .8 mm., 1.3 mm. and 1.4 mm.

Pedicle valve highly convex, convexity being greatest in the umbonal region; beak incurved. The reduced convexity on the postero-lateral margins in the small specimen gives the appearance of a slight depression just in front of the hinge-line and in the large specimens of a flattened area. A low, rounded elevation is not infrequently found in the anterior area of the larger specimens. Cardinal area low, slightly arched, vertically striate; hinge-line denticulate. Deltidium small, higher than wide and elevated above the area. Two heavy, diverging ridges bound the muscular scars laterally, and anteriorly they are highly elevated, making the scars appear deep and spoon-shaped. Beneath the beak are two crests from which a short, rather high, angular median septum extends anteriorly. On either side of the median septum and just below the crests are the deep subelliptical adductor scars. Genitalial markings consist of pustules over all the surface from the muscular scars almost to the margins of the shell.

The brachial valve concave, with the greatest concavity near the middle. Postero-lateral slopes flattened. Cardinal area low, vertically striate. Hinge-line crenulate. Chilidium higher than wide, elevated above the area and slightly flattened. From the base of the very prominent bifid cardinal processes originates a very short, heavy, rounded median septum which bifurcates to form the lateral boundaries of the anterior adductor scars. On either side of the bifurcated median septum are protuberances representing the crural plates. Lateral slopes pustulose; margins marked by fine granules.

Surface of both valves marked by moderately heavy rounded costae separated by from 3 to 8 finer, somewhat irregular costae. The heavy ones increase by implantation and the finer by both bifurcation and implantation.

Remarks. — The species *arcuata* appears to be by far the commoner *Douvillina* of the Hackberry. It appears in the Gypidula faunule, is abundant throughout the Spirifer zone, and disappears in the Owen. Whether or not the identifications of *D. arcuata* from other parts of the country are reliable is uncertain; we have emphasized, in discussion of other species and in the section on stratigraphy, the essential uncertainty of most of such identifications.

Occurrence. — *D. arcuata* appears in the Gypidula faunule, and is abundant in the Spirifer zone, but becomes rare in the Owen, finally disappearing.

Plesiotypes. — Nos. 7854 to 7857, University of Michigan.

DOUVİLLINA MAXIMA C. L. Fenton
(Plate XX, Figs. 11–16)

Douvillina arcuata maxima C. L. Fenton, Am. Mid. Nat., vol. 5, p. 215 pl. 6, figs. 23–25. 1918.

Description. — The general shape of the small specimens is quite similar to that of *D. arcuata*, but the mature individuals are more nearly quadrangular. Greatest width of the shell near the mid-length and the hinge-line less than the greatest width. This is in contrast to *D. arcuata* which, in most cases, has its greatest width along the hinge-line. Dimensions of the larger cotype and the plesiotype: length of pedicle valve, 21.5 mm. and 14 mm.; length of brachial valve, 18.8 mm. and 13 mm.; width, 25.4 mm. and 17.3 mm.; length of hinge-line, 22.2 mm. and 16.9 mm.; height of cardinal area, 2 mm. and 1.5 mm.; thickness, 11 mm. and 7 mm. The interior of the pedicle valves shows a continuation of the median septum beyond the muscular scars. Genitalial areas flabelliform with the markings in the form of fine irregular ridges or pustules. Fine ridges on the anterior margins mark the vascular areas. The brachial valve of a mature heavy individual shows the same characters as those in *D. arcuata*, though it is much coarser.

Surface of both valves marked by rather heavy costae separated near the beak by from 1 to 3 finer ones that, in most cases, become heavier as they progress anteriorly. Fine ones are implanted so that at the anterior margin the heavy ones are separated by from 2 to 4. Shell punctate.

Remarks. — The cotypes, originally described as a variety of *D. arcuata*, represent the old age forms of the species *maxima*, which is not common. The surface markings, which are characteristic, are much worn on these specimens.

Occurrence. — *D. maxima* appears in the Gypidula faunule and continues throughout the Spirifer zone and into the Owen. It is most common in the Brachiopod faunule at Rockford.

Cotypes. — No. 26057, Walker Museum; *Plesiotypes.* — Nos. 8016 to 8020, University of Michigan.

DOUVILLINA DELICATA, n. sp.

(Plate XX, Figs. 17–20)

Description. — Shell below medium size; semicircular in outline, broader than long with the greatest width near the mid-length; posteriorly there is an incurving of the margin which gives the hinge-line a mucronate appearance. Dimensions of two typical specimens: length of shell, 19.5 mm. and 16.9 mm.; width, 21.9 mm. and 19.7 mm.; thickness, 5.2 mm. and 3.4 mm.; height of cardinal area, 1.8 mm. and 1.3 mm.

Pedicle valve moderately convex; mesial fold low and rounded, from 5 to 7 mm. in width at the anterior margin. Cardinal area horizontal and moderately low; marked by fine vertical striae, which are crossed by coarse horizontal lines. Delthyrium flat, triangular, marked by fine growth lamellae. Brachial valve moderately concave; mesial sinus shallow and very indistinctly bounded. Cardinal area linear and vertical; marked as is that of the opposite valve. Surface of both valves marked by coarse angular costae. The costae are crossed by very fine irregular concentric striae which form nodes on the costae. The surface, particularly in the umbonal region, appears to be punctate.

Remarks. — This species differs markedly from the other American forms of this genus in size, degree of convexity, and coarseness of plications. The internal characters, however, serve to establish the genus.

Occurrence. — Upper beds of the Spirifer zone, particularly at Bird Hill. It is nowhere common.

Holotype. — No. 7942; *Allotype.* — No. 7943; *Paratype.* — No. 7944, University of Michigan.

GENUS LEPTOSTROPHIA HALL & CLARKE

Leptostrophia Hall and Clarke, Pal. N. Y., vol. 8, pt. 1, pp. 287–288. 1892
Leptostrophia Hall and Clarke, Eleventh Ann. Rep. N. Y. State Geologist,
    p. 281. 1894.

Description. — This genus, described under *Stropheodonta* by
the authors, has never received very full definition. The
original description [16] follows:

"The plano-convex species of *Stropheodonta* are distinguished
from the group of *S. demissa* by more than contour alone. The
characters of the deltidium show the same progressive development as in the concavo-convex Stropheodontas, the earliest species
having the delthyrium sometimes open, sometimes partially
closed by a convex plate; while in. the Devonian species the
deltidium is reduced to a flat, transverse lamina, supported
within by the callosity about the cardinal apophyses. In the
pedicle valve are two very strongly pustulose, diverging ridges,
bounding the muscular impressions on their lateral margins,
while anteriorly these scars are broadly flabelliform and not
strongly limited. The central adductors are small, relatively
obscure, and not divisible."

The Hackberry species here referred to the genus *Leptostrophia* show clearly the internal features described above. The
character of the delthyrial region, however, is markedly different.
Instead of having a "flat, transverse lamina," *S. rockfordensis*
and its allies possess two short, yet moderately broad, deltidial
plates which, in all specimens examined, fail to reach the hinge-
line. Between these plates, and apparently of later origin, is a
third one, similar to the delthyrial plate in the Telotremate
genera *Pseudosyrinx* and *Platyrachella*, for which we propose the
name 'pseudo-delthyrial plate.' Inasmuch as all of the specimens of *Leptostrophia magnifica* and other typical species from

[16] *Pal. N. Y.*, vol. 8, pt. 1, p. 287.

the East and from Missouri that we have examined do not have the area well preserved, we cannot say whether or not this plate combination is typical of the genus. However, should it prove that the Iowa forms deserve separate designation, either as a genus or subgenus, the characters of the cardinal region will prove to be the diagnostic feature, along with greater convexity.

LEPTOSTROPHIA ROCKFORDENSIS, n. sp.

(Plate XXI, Figs. 3-7)

Stropheodonta perplana nervosa Calvin, Ia. Geol. Surv., vol. 7, p. 167. 1897.
Leptostrophia perplana nervosa, C. L. Fenton, Am. Journ. Sci., 4th ser., vol. 48, p. 371. 1919.

Description. — Shell above medium size to large, semicircular in outline, concavo-convex, broader than long with the greatest

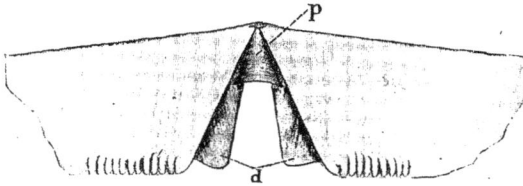

FIG. 6. Cardinal Area of *Leptostrophia rockfordensis*, showing the Deltidia Plates (d) and Pseudo-delthyrial Plate (p). × 10.

width near the mid-length. Hinge-line less than the greatest width, cardinal extremities rounded or obtusely angular. Dimensions of the holotype: length of pedicle valve, 30.3 mm.; width of pedicle valve, 42 mm.; length of hinge-line (extremities broken off), 37.3 mm.; height of cardinal area, 2.7 mm.; thickness, 4.6 mm. (See Fig. 6.)

Pedicle valve slightly convex with greatest convexity in the umbonal region, which, however, in some specimens is considerably flattened in the vicinity of the beak. Lateral margins flat;

anterior margin slightly convex; holotype considerably distorted around the anterior margin. Cardinal area quite low, inclined posteriorly without arching; marked by fine horizontal costae and vertical lines, about six in the space of 2 mm. These are very prominent when the shell is weathered and seem to correspond to the hinge-teeth. Delthyrium large; length and width about equal. Two pustulose, heavy, diverging ridges bound the lateral margins of the muscular impressions; anteriorly the scars become broadly flabelliform and are indistinctly bounded. The adductor scar linear, on a median ridge dividing the muscular area; in one case this ridge divides near the apex and the adductor scar is located between the two parts. Genitalial markings, consisting of pustules of varying size, are conspicuous in the area between the diductor scars and the posterior margin.

Brachial valve flat or slightly convex; in specimens at hand it is neither as wide nor long as the pedicle valve, but this is due to the delicacy of the marginal areas. Cardinal area very narrow, forming an obtuse angle with that of the opposite valve; marked by vertical lines that probably correspond to the denticulations. Two lateral diverging muscular ridges strongly pustulose; anterior to these and near the center of the shell are two pustulose muscular elevations. Pustulose genitalial markings on the lateral margins.

Surface of pedicle valve marked by coarse nodose irregular costae. Finer markings between the costae in the umbonal region become coarse and nodose anteriorly. In the anterior margin costae are implanted. Nodes are very angular, rather long, high and almost forming spines, especially near the front of the shell. Lines of growth heavy, irregular and lamellose, particularly near the margin. The brachial valve is marked by fine radiating striae with from 3 to 5 still finer ones between them. Concentric lines of growth very fine, irregular and lamellose.

Remarks. — This is the species that has usually been identified as *L. perplana*, or its variety *nervosa*. From both of those forms it differs in greater thickness and coarseness of shell, coarser costae irregularly arranged, and coarser internal markings.

Occurrence. — Throughout the Spirifer zone, particularly common in the Strophonella faunule at Bird Hill. Common also in the Gypidula faunule below the Spirifer beds.

Holotype. — No. 26058, Walker Museum; *Allotype.* — No. 7934; *Paratypes.* — No. 7937, University of Michigan.

LEPTOSTROPHIA CANACE (Hall & Whitfield)

(Plate XXI, Figs. 1–2)

Strophodonta canace Hall and Whitfield, 23d Ann. Rep. N. Y. State Cab. Nat. Hist., p. 236, pl. 11, figs. 8–11. 1873.

Description. — Shell medium size or smaller, concavo-convex, broader than long; extended hinge-line. Dimensions of the two plesiotypes: length of pedicle valve, 14 mm. and 18.4 mm.; length of brachial valve of the second specimen, 16.9 mm.; width at mid-length, 19 mm. and 25.3 mm.; width along hinge-line (the second specimen is broken), 29.9 mm. and 21.5 mm.; thickness of the second specimen, 6.7 mm.; height of the cardinal area of the second specimen, 1.5 mm.

Pedicle valve semicircular with the greatest convexity near the mid-length of the shell. Beak elevated slightly above the general surface, but not prominently; surface slopes gradually from the umbonal region to about the middle, where it either abruptly or gradually slopes to the anterior margin; postero-lateral area usually flattened. Cardinal area approximately horizontal, marked by vertical striae; denticulate. Delthyrium small, about as wide as high, and closed by the deltidial and pseudo-delthyrial plates like those of *L. rockfordensis;* but in this species the pseudo-delthyrial plate is enlarged so as partially or wholly to obscure the deltidial plates.

Brachial valve plano-concave. Cardinal area linear and denticulate. The chilidium is highly convex and extends slightly beyond the brachial valve into the delthyrium.

Surface of pedicle valve marked by heavy rounded costae that increase in number by implantation, 5 to 8 in the space of 5 mm.; between these are 2 to 9 fine wavy costae that increase by separation from the boundaries of the larger costae. Fine concentric lines mark the entire surface. Surface of the brachial valve marked by low, rounded, wavy costae, which are crossed by concentric lines. Very irregular growth lamellae mark the surface of both valves of one of the plesiotypes.

Remarks. — *L. canace* is distinguished from *L. rockfordensis* by its smaller size, greater convexity and different type of surface markings.

Occurrence. — Spirifer zone, particularly the Leptostrophia faunule.

Plesiotypes. — Nos. 7930 and 7931, University of Michigan.

LEPTOSTROPHIA CAMERATA, n. sp.

(Plate XXI, Figs. 8–13)

Stropheodonta calvini Calvin, Ia. Geol. Surv., vol. 7, p. 167. 1897.

Description. — Shell small, subquadrangular in outline, concavo-convex, broader than long; hinge-line produced. Dimensions of holotype with extended hinge-line and allotype: length of pedicle valve, 10.6 mm. and 19.4 mm.; length of brachial valve of the allotype, 8.5 mm.; width at the mid-length of shell, 13.8 mm. and 13.5 mm.; width along hinge-line, 22 mm. and 12.7 mm.; thickness of allotype, 4.7 mm.

Pedicle valve convex with the greatest convexity near the

mid-length of the shell; umbonal area flattened. Mesial sinus originates a little anterior to the mid-length and develops into a broad flat area anteriorly; that of the holotype measures 5 mm. at the margin of the shell. Cardinal area very low, uniform throughout, flat and vertically striate; hinge-line denticulate. Muscular scars bounded laterally by two heavy diverging ridges for about 2 mm., then by much smaller ones, on some specimens very indistinct; anteriorly the scars are not bounded. Beneath the beak are two crests from which a rather heavy elevated median septum extends anteriorly; from the sides of these crests originate low, rounded ridges that parallel the heavy ones and bound the oval-shaped adductor scars laterally. Genitalial markings of pustules on the lateral areas.

Brachial valve plano-concave with the greatest concavity near or anterior to the mid-length of the shell; usually smaller than the pedicle valve. Cardinal area linear.

Surface of the pedicle valve marked by heavy rounded costae that increase in number by implantation; 6 to 7 in 5 mm. at the anterior margin. On some specimens the heavy costae are irregular and have angular nodes that almost form spines; 3 to 9 fine costae between the heavy ones. Surface of brachial valve marked by low rounded ridges — 7 in 5 mm. about mid-length of the shell — the crests of which usually bear one large, rounded costa. Fine, and often wavy, costae mark the surface.

Occurrence. — Spirifer zone; particularly abundant in the Leptostrophia faunule.

Holotype. — No. 7938; *Allotype.* — No. 7939; *Paratypes.* — Nos. 7940, 7941, and 8145, University of Michigan.

GENUS STROPHONELLA HALL

STROPHONELLA REVERSA Hall

(Plate XXII, Figs. 1–17)

Strophodonta reversa Hall, Geol. Ia., vol. 1, pt. 2, p. 494, pl. 3, fig. 4. 1858.
Strophonella reversa Hall, 28th Ann. Rep. N. Y. State Cab. Nat. Hist.,
    p. 154. 1879.
Strophonella reversa Hall and Clarke, Pal. N. Y., vol. 8, pt. 1, p. 293, pl. 12,
    figs. 16–20. 1892. (See remarks.)

Description. — Shell of medium or more than medium size,
wider than long, with the hinge-line considerably produced beyond the postero-lateral margins. Dimensions of five typical
specimens: length of pedicle valve, 5 mm., 9.7 mm., 13.4 mm.,
19.9 mm. and 21.1 mm.; length of brachial, 4.3 mm., 8.6 mm.,
11.3 mm., 17.6 mm. and 20.2 mm.; width along hinge-line, 7.8
mm., 15.5 mm., 22.9 mm., 31.9 mm. and 34 mm.; width at
mid-length of valve, 6.1 mm., 11.6 mm., 15 mm., 22.4 mm. and
25.3 mm.; thickness (= convexity), 2 mm., 3.1 mm., 3.2 mm.,
4.8 mm. and 9 mm. Of the last measurement, the brachial
valve consumes about 7 mm. while the remaining 2 mm. represent the elevation of the pedicle beak above the plane of the
shell.

Pedicle valve convex throughout the neanic stages of the
shell, with the umbo high and prominent, beak sharp and incurved, lateral and anterior slopes moderately convex, and the
cardinal extremities sharply produced. This stage of development is typically represented in the smallest of the specimens
measured and illustrated. Somewhat farther on in development
the lateral slopes become flattened, the umbo less elevated, the
cardinal area broader and flatter, and the beak little more than
a slightly upward-curved point at the center of the area. The
cardinal extremities are still sharply extended and differentiated
from the postero-lateral slopes (Figs. 3–4). In the next stages

(Figs. 5–6) the flattening of the slopes and the area continues, while the valve as a whole becomes less convex. The sharp curve from postero-lateral margins to cardinal extensions is replaced by a very gradual one, and in the more advanced specimens the anterior portions begin to show concavity, while the postero-lateral portions and the productions are decidedly flattened. Such a specimen as that illustrated in Figures 7–8 may well be said to represent the youth of the species.

Maturity is best represented by the specimen shown in Figures 14–15. In it the pedicle valve is prominent and convex, with the beak pointed, but only moderately incurved where it projects beyond the area. The area is broad, flattened and about 2.5 mm. high; the delthyrium, as in all of the other stages described, is completely closed by the deltidium, which is broadly triangular and flat, being distinguished from the area by smoothness and darker color. Area marked by fine, vertical striae. Postero-lateral slopes flattened; antero-lateral and anterior ones convex. Mesial portion of valve marked by a strong ridge.

The development of the brachial valve corresponds closely to that of the pedicle, though it is, of course, reversed. In the early neanic stages it is strongly concave, with flattened, upturned cardinal extensions and an almost linear area. With the progress of growth, the valves become less concave, the postero-lateral slopes more flattened, the area higher, and finally, in youth, the anterior portion of the shell becomes convex. From youth to maturity there are no striking changes; the valve develops a greater convexity and the cardinal extremities become less acute, but distinctly new developments are lacking. In the typical *reversa* adults, the mesial depression continues to the margin, even being accentuated in the more convex individuals.

The entire surface of the valves is punctate. The markings consist of striae which in the neanic stages are simple, strong, sharply angular, and separated by broad, flat furrows. Within 4 mm. of the beak their number is increased by implantation; within 9 or 10 mm., dichotomizing begins and is repeated two or three times; trifurcation as well as bifurcation is common. Because of the increase in number and division, the striae at the anterior slopes are much finer than those near the beak. The cardinal extremities commonly are marked by diagonally directed wrinkles, ·such as are found in various species of *Strophomena*. Teeth prominent on cardinal margins.

Interiors of pedicle valves belonging to specimens in the youthful stage of development show the umbonal region deeply concave, with the diductor scars very faintly bounded. In all specimens, however, there is a distinct separation between the scars of the anterior and posterior diductors, the latter being considerably deeper. The adductor scars are indistinguishable; the cardinal process is prominent and bilobed, and on either side are large depressions for the reception of the processes of the opposite valve. Numerous radially directed vascular markings reach the anterior margin.

Pedicle valves of mature specimens show the diductor areas large and leaf-shaped, with elevated borders, while the adductor scars are quite distinct. About 3 mm. posterior to the margin runs a high, heavy ridge, which is bisected by a deep furrow which corresponds to the elevated mesial ridge of the exterior. Vascular markings are present, as in the youthful specimens, but are stronger and more irregular. Over the vascular areas are scattered the tuberculate genital markings.

Brachial valves in the youthful stage of development have the interior flattened or even slightly concave. The cardinal processes are long, and sharply curved posteriorly. The anterior and posterior adductor scars are distinct, and the vascular area is defined by a low, irregular thickening of the shell, which is crossed by vascular markings which extend to the margins.

Anterior to the muscular area there originates a broad, rounded or subangular septum which reaches its maximum height of 2 or 3 mm. posterior to the elevation surrounding the vascular area, and continues to the margin as a low, broad ridge bearing vascular channels.

Mature valves resemble the young ones in arrangement of markings only. They are deeply convex, with high, heavy cardinal processes and strong muscle scars. The vascular area is deeply concave, and marked by numerous coarse pustules. The ridge-like boundary of the vascular area is high and heavy, and is crossed by numerous deep channels which ramify variously before they reach the margin. The septum is very large and heavy.

Remarks. — This highly variable and extremely characteristic Hackberry form was described and illustrated by Hall, apparently from rather meager material. The specimens used by Hall and Clarke may or may not have properly represented the species. The specimen shown in Figures 16 and 17 of their Plate XII appears to represent with considerable fidelity a specimen of *reversa* that is somewhat longer and heavier than the form which is here considered as being the typical species. Figure 18, however, appears to be of a specimen which, though it may be a *reversa*, is nearer to the species *hybrida* as it is here defined. Figure 19 resembles the large, heavy type here referred to as *ponderosa;* what Figure 20 may represent is uncertain. It does, however, give an excellent, though somewhat diagrammatic, picture of the typical condition of the areas and deltidia in all of these forms.

In attempting to re-define Hall's *reversa* we have selected the youngest specimens which show the specific character, and have followed the development from the neanic to the mature stages. For the mature types we have selected those individuals which show the closest possible correspondence with the youthful forms but which, in their size, thickness and general proportions, give positive evidence of being mature. It is on the basis of this type, which corresponds closely to the one illustrated by Hall, that we shall define the several variations exhibited by the species.

Occurrence. — *Strophonella reversa* ranges throughout the Spirifer zone of the Hackberry, being abundant in all localities where that division is exposed. The typical form is found in abundance at Rockford, Hackberry and Bird Hill, and is particularly common in the Brachiopod faunule and its equivalent. Because of preservation, the young are more common at Bird Hill and Rockford than at Hackberry Grove and in the hills along Lime Creek.

Whether or not this species occurs in other formations we cannot say. It has been identified from the Naples beds in New York, by Schuchert, but we have not examined specimens from any formation but the Hackberry.

Plesiotypes. — Nos. 8002 to 8010, University of Michigan.

Variations of Strophonella reversa

As may be seen from any series that contains a dozen or more specimens, *Strophonella reversa* is a highly variable species. Certain of those variations are easily distinguishable, and may be identified with but little difficulty, while others are poorly defined and of slight permanence. None of them, so far as present collections show, possess any particular stratigraphic significance. For purposes of description, these variations are grouped under two headings, minor and major, the former being discussed first.

STROPHONELLA REVERSA, form ALTA nov.

(Plate XXII, Figs. 20–22)

Strophonella reversa of most identifications.

Description. — Shell of medium or less than medium size, as wide as, or wider than, long, with the greatest width along the hinge-line. Dimensions of three typical specimens: length of pedicle valve, 16 mm., 22.1 mm. and 22.4 mm.; length of brachial valve, 15.6 mm., 20.9 mm. and 21.2 mm.; width, 18.6 mm., 24.7 mm. and 26.5 mm.; thickness (= convexity), 5.6 mm., 8.5 mm. and 10.2 mm.

In its ontogeny this form conforms to *S. reversa* till the stage of youth. From this point on, however, there is a distinct change in development. Instead of continuing to build up the wide shell, with long cardinal extensions, those animals of the *alta* line directed their energies to the building of extensions along the frontal margin. As a result of this, the shells are decidedly subelliptical, the length almost equaling the width. Mucronations of the hinge-line are very short, or altogether lacking. The thickness of the shell, as well as its convexity, is greater than that of the typical *reversa*.

Remarks. — This form, which is closely associated with *S. reversa*, can hardly be considered a mere old-age or gerontic development. Aged individuals of *reversa* tend toward high convexity and greater elongation, it is true, but those developments do not begin so early, nor reach such an extent as in *alta*.

Occurrence. — Associated with *Strophonella reversa* wherever that species is found, but particularly common in the middle portion of the Spirifer zone at Rockford and Hackberry Grove.

Holotype. — No. 8100; *Allotype.* — No. 8101, University of Michigan.

STROPHONELLA REVERSA, form LATA nov.

(Plate XXII, Figs. 18–19)

Strophonella reversa of most identifications; *S. hybrida* of some.

Description. — Shell of varying size, most commonly less than medium; width greater than length, with greatest width along the hinge-line. Dimensions of three typical specimens: length of pedicle valve, 15 mm., 18.9 mm. and 23.3 mm.; length of brachial valve, 12.8 mm., 17.8 mm. and 21.5 mm.; width, 17.3 mm., 23.6 mm. and 29.2 mm.; thickness, 5.2 mm., 7.7 mm. and 9.4 mm.

This form appears to occupy an intermediate position between *S. reversa* and *S. hybrida*. Some of the individuals, however, show decided *reversa* relationships, and they have been selected as the types of this form. The smallest of these is a heavy, compact shell, with plications which show typical *reversa* bifurcation, and which has a very wide cardinal area, ribbed as in *reversa*. In shape the shell differs markedly from that species, being almost rectangular in general outline. Thus for the greatest portion of their length, the lateral margins curve so gradually that 9.2 mm. anterior to the cardinal margin of the brachial valve the width of the shell is 16.1 mm. — only 1.2 mm. less than that at the hinge-line.

The second specimen is somewhat closer to the typical *reversa* than is the first, manifesting considerable similarity to *S. reversa alta*, but shows the straightened lateral margins of the typical *lata*. The third and largest specimen is much like the first, though the proportion of length to width is very slightly less in the larger of the two. The difference, however, is so slight as to be unimportant.

Remarks. — This form is closely related, or appears to be closely related, to *Strophonella hybrida*, and is most commonly found in the beds where that species is abundant. The relationship to *S. reversa* is shown by the tendency toward extension of the hinge-line, the small size of most specimens, the greater convexity of the brachial valve during youth, and the long shell which not uncommonly possesses a shape akin to that of *S. reversa alta*. It was this form which caused one of the present authors to place the species *hybrida* as a variety of *reversa*.

Occurrence. — Throughout the Spirifer zone of the Hackberry, especially in the Strophonella faunule and the corresponding beds as Hackberry and Bird Hill.

Holotype. — No. 8015; *Allotype.* — No. 8014, University of Michigan.

MISCELLANEOUS MINOR VARIATIONS

(Plate XXII, Figs. 16–17)

Associated with the several readily distinguishable forms of *S. reversa* are numerous aberrant types which probably represent nothing more than individual variations. In most cases they are associated with old age; in others injury may be the principal cause. In the main these variations are manifest in elongations of the shell, excessive thickening and convexity in the anterior portion, and in extreme thickening of the whole shell.

MAJOR VARIATIONS OF STROPHONELLA REVERSA

STROPHONELLA REVERSA GRAVIS C. L. Fenton

(Plate XXII, Figs. 23–30)

Strophonella reversa gravis C. L. Fenton, Am. Mid. Nat., vol. 5, p. 214, pl. 6, figs. 18–22. 1918.

Description. — Shell of medium size, with the length almost equal to the width, and the hinge-line not produced, but instead commonly shorter than the maximum width of the valve. Di-

mensions of three specimens, the first and third being the
cotypes: length of pedicle valve, 23 mm., 24.5 mm. and 25.7
mm.; length of brachial valve, 22.9 mm., 23.3 mm. and 25 mm.;
width, 23.6 mm., 26.4 mm. and 27.7 mm.; thickness, 10.1 mm.,
12.6 mm. and 12.4 mm. In the last two specimens the greatest
width is almost 10 mm. anterior to the hinge-line.

Pedicle valve elevated and convex in the umbonal region,
with a prominent though obtuse beak, broad, striated cardinal
area, and distinct mesial elevation which extends to the anterior
margin. Anterior and lateral portions of the valve concave;
cardinal regions flattened. Interiorly the valve is much like a
diminutive and highly exaggerated *S. hybrida ponderosa.* The
bifid cardinal process is broad and heavy, and the sockets for
the process of the opposite valve are deep and large. The
diductor scars are large, deep, and elevated at the margins,
with the scars for the anterior and posterior muscles sharply
distinguished. Adductor scars broad and deep. Vascular region
narrow and semicircular, marked by heavy pustules. About the
vascular region, and from 2 to 5 mm. inward from the margin,
runs a broad, heavy ridge which increases anteriorly and is cut
at the median line by a deep channel which originates between
the diductor scars and extends to the margin, its course corresponding to that of the external mesial elevation. The ridge is
crossed by irregular, ramifying channels which become indistinct
on the marginal slope of the ridge.

Brachial valve concave near the umbo, convex near the
middle, and flattened anteriorly. Mesial depression distinct in
the umbo, but faint or lacking on the anterior slope. Interior
very heavily and coarsely marked. The adductor scars are deep,
but of various sizes, and are situated on a broad, heavy, calcareous elevation. Upon this elevation is an elongate ridge,
the scar of the diductors. Anteriorly the elevation merges into
a strong median septum which reaches its maximum height near

the mid-length of the valve, but which continues to the margin. In the gerontic specimens it is marked by strong, deep, vascular channels. Vascular areas are deep pits, for the reception of the strong prominences of the opposite valve; about these pits runs a broadly rounded elevation, which is crossed by the ramifying vascular channels. Genital markings consist of coarse pustules which are found not only in the vascular depressions, but on the septum and muscular elevation as well. The cardinal processes are long and curved, and the hinge-teeth prominent, as are those of the pedicle valve.

Remarks. — The prime distinctive feature of this shell is its great heaviness. Some, it is true, fail to show this character very clearly, but by their length and form indicate the variety to which they belong. That the variation is not an incidental affair, due to old age alone, is shown by the small size of many of the specimens, one very heavy pedicle valve measuring 10.8 mm. in length and 15.2 mm. in width. The only question is as to the propriety of referring all of the heavy individuals to the form *gravis*. But in spite of the obvious variation in shape, it was found to be impossible to make any separation of even minor importance.

Occurrence. — *S. reversa gravis* is fairly common throughout the Spirifer beds, but is most common in the portion below the Leptostrophia faunule. In the uppermost beds, such as the Strophonella faunule at Bird Hill, it is relatively uncommon.

Cotypes. — Nos. 26059 and 26060, Walker Museum; *Plesiotypes.* — Nos. 8092 to 8097, University of Michigan.

STROPHONELLA REVERSA TRIANGULARIS, n. var.

(Plate XXIII, Figs. 9–11)

Description. — Shell larger than medium, wider than long, with the greatest width along the hinge-line. Dimensions of the holotype: length of pedicle valve, 25.5 mm.; length of brachial valve, 22.6 mm.; width, 34.7 mm.; thickness, 9.3 mm. Height of cardinal areas 3.6 mm. and 2.5 mm. respectively in the holotype and allotype.

In general type and ontogeny this variety resembles *S. reversa.* However, it is considerably larger and heavier, has more distinct wrinkles on the cardinal slopes, and higher cardinal areas. On the pedicle valve the mesial ridge gives way to a strong, upwardly directed depression, somewhat like that to be found in *Strophomena nutans* of the Ordovician; this character is shown particularly well in the allotype. On the brachial valve the mesial depression gives rise to an elevation. The moderately curved lateral margins give the shell a distinctly trapezoidal appearance. The interior of the shell shows no specially distinctive features.

Occurrence. — This variety has been found only in the Spirifer beds, at Bird Hill, Rockford and Hackberry Grove.

Holotype. — No. 8098; *Allotype.* — No. 8099, University of Michigan.

STROPHONELLA OBESA, n. sp.

(Plate XXIII, Figs. 4–7)

Description. — Shell large, of medium size, wider than long, and with greatest width anterior to the hinge-line. Dimensions of the holotype and allotype: length of pedicle valve, 19.6 mm. and 25.1 mm.; length of brachial valve, 18.7 mm. and 25 mm.; width, 25 mm. and 31.9 mm.; thickness, 8.4 mm. and 10.5 mm.

Pedicle valve moderately convex in the neanic stages; convex with flattened or concave postero-lateral slopes in youth, and convex umbonally, but deeply concave both anteriorly and laterally in mature stages. The neanic stages, as indicated by growth lines, fail to show the long cardinal extensions typical of *S. reversa;* in the adults the cardinal extremities are rounded or obtuse. The one pedicle interior in our collections shows the diductor scars to be deep and strongly ridged, and elevated as much as 2.2 mm. at the margins. The vascular area is broad and flattened and marked by coarse genital pustules. The extravascular ridge is low and irregular, the mesial depression indistinct, and the vascular sinuses shallow.

The pedicle valve is deeply concave near the beak in the allotype and but slightly so in the holotype. The maximum convexity is found about mid-length of the valve, and the anterior and lateral slopes are moderately convex or slightly flattened.

Surface of shell marked by high, coarse, acutely angular costae which increase both by implantation and division. Near the brachial beak of the holotype there are 4 or 5 of these plications in the space of 3 mm.; near the margin the number is about 7 in the same space.

Remarks. — This species may be distinguished by its very strong, sharply angular costae, great convexity, and lack of cardinal extensions. In all of these characters it differs markedly from *Strophonella reversa* and *S. hybrida,* and all of their varieties.

Occurrence. — This species appears to be confined to the lower portions of the Spirifer zone, and to the Gypidula beds below. Specimens have been found at Rockford and Hackberry Grove.

Holotype. — No. 8124; *Paratype.* — No. 8126, University of Michigan; *Allotype.* — No. 26061, Walker Museum.

STROPHONELLA OBESA PLICATELLA, n. var.

(Plate XXIII, Fig. 8)

Description. — This form is known from a single damaged specimen, the dimensions of which are: length of pedicle valve, 23.2 mm.; length of brachial valve, 22.8 mm.; width, about 32 mm.; thickness, 7.6 mm. In general appearance the specimen resembles the allotype of *S. obesa*, but is much thinner and less convex, and possesses decided wrinkles in the region of the cardinal extremities. These, too, are slightly produced, so that the maximum width is along the hinge-line.

The most distinctive feature, and the one on which the variety is based, is the type of the radiating costae. Where those of *obesa* are large, coarse and angular, those of *plicatella* are fine, regularly disposed and rounded. The increase, as in *obesa*, is both by implantation and bifurcation, but the latter is the more prevalent. Also, there is but little increase on the anterior third of the shell, so that at the margin the costae are more loosely spaced than they are in the region of maximum convexity. Thus at 5 mm. from the brachial beak there are 6 costae in 3 mm.; at 15 mm., from 7 to 9, with 8 as the commonest number; and at the margin, 5, 6, and rarely 7 in 3 mm. The difference between the two forms may readily be seen by comparing the illustrations of the two holotypes.

Occurrence. — The holotype was collected in the lower beds of the Strophonella faunule, at Bird Hill.

Holotype. — No. 8127, University of Michigan.

STROPHONELLA HYBRIDA (Hall and Whitfield)

(Plate XXIII, Figs. 12–15; Plate XXIV, Figs. 1–10)

Strophodonta hybrida Hall and Whitfield, 23d Ann. Rep. N. Y. State Cab.
Nat. Hist., p. 239. 1873.
Strophodonta reversa Schuchert, Bull. U.S. G. S., vol. 87, p. 439. 1897.
Strophonella hybrida Calvin, Ia. Geol. Surv., vol. 7, p. 167. 1897.
Strophonella reversa hybrida C. L. Fenton, Am. Journ. Sci., 4th ser., vol. 48,
p. 371. 1919.

Description. — Adult shells of large size, wider than long,
with the greatest width along the hinge-line. Dimensions of
three specimens of late neanic, youthful and mature stages
respectively: length of pedicle valve, 14 mm., 21.6 mm. and
23.7 mm.; length of brachial valve, 12.7 mm., 19 mm. and
25.9 mm.; width, 20.1 mm., 29.5 mm. and 36 mm.; thickness,
3.9 mm., 5 mm. and 7.6 mm.

The ontogeny of *S. hybrida* follows closely that of *S. reversa*
through the neanic stages; indeed, the younger neanic specimens
of the two species are indistinguishable. Differentiation appears
when the shells are 12 to 26 mm. in width; those of *hybrida* are
distinguished by lesser concavity of the brachial valve, slightly
finer plications, and a lesser development of the cardinal extensions. The youthful *hybrida* differs from the *reversa* of the same
stage in these features, as well as in greater size and lesser
concavity of the pedicle valve and in the convexity of the
brachial. The interiors, too, fail to show any very marked
differences from those of *reversa* in late youth or very early
maturity.

Such, then, is the traceable ontogeny of the species. The
remainder of this description applies only to adult specimens.

Pedicle valve moderately convex near the umbo, concave
anteriorly, and flattened on the postero-lateral portions, along
the cardinal margin. Beak short, obtuse, and very slightly

incurved; cardinal area wide and high, vertically striate, with the delthyrium completely closed by a flat, triangular deltidium, which is quite smooth. Many specimens show the cardinal extremities wrinkled, while in others they are quite smooth. Interior of the valve shows the cardinal margin to be marked by teeth almost to the extremities; the cardinal process low, heavy and slightly bifid. Diductor scars broadly subovate or more or less subovate in some specimens, but in most of them the scars are narrowly and irregularly flabellate, bounded by high, irregular ridges, and separated into posterior and anterior scars. The scar of the adductors lies between those of the diductors. Vascular area smooth or marked by vascular channels and in some specimens, by genital pustules. About 5 mm. from the margin, and bounding the vascular area, runs a ridge which is either rounded or subangular, and which is truncated by the mesial depression. On the marginal slopes of the shell the vascular channels ramify and bifurcate or trifurcate once or twice before reaching the margin.

Brachial valve moderately concave in the umbonal region and on the cardinal extremities, convex anteriorly, and with or without a mesial depression. This is present only in those specimens which show a corresponding elevation on the opposite valve. Extremities commonly crenulate. Interior of valve shows the cardinal processes to be long and heavy, the adductor scars irregular and elevated, the diductor scar linear, the vascular area broadly subcordate, marked by genital pits and bounded by a broad, low ridge or thickening of the shell. It is divided medially by a heavy, low median septum which may or may not continue to the margin. Vascular channels originate within the vascular area and extend to the margin.

Surface of the shell punctate; marked by large numbers of coarse or moderately fine radiating costae, about 7 or 8 of which occupy the space of 5 mm. These costae increase by bifurcation and implantation, and are of variable strength. Most commonly there is one finer costa between two heavier ones, but the arrangement is more or less irregular. The costae are crossed by coarse or fine concentric growth lines which originate within 5 mm. of the beaks and increase in number and distinctness toward the margin.

Remarks. — Although Hall and Whitfield did not illustrate their description, there can be little question that this is the species to which they referred. *S. hybrida* is distinguished from *S. reversa* by its greater width, lighter shell, less acute extremities, higher cardinal areas, finer costae and much greater size. As has been pointed out in the discussion of the minor variations of the species *reversa*, there are forms which are, or appear to be, intermediate between the two, and these led C. L. Fenton to express the opinion that *hybrida* was but a variety of *reversa*. Similar specimens, probably, induced Schuchert to place *hybrida* as synonymous with the smaller species.

It will be noted that among the illustrations there are figures of interiors of pedicle valves that appear to be more or less intermediate between the true *hybrida* and its variety *ponderosa*. These valves, because of their thickness and strength are by far the commonest of the *hybrida* remains, but because of the lack of brachial valves which give unmistakable evidence of corresponding to them, it has been impossible to determine their relationship certainly. Probably, however, they are gerontic developments of *hybrida;* less probably, intermediate stages between the species and its variety.

Occurrence. — Throughout the Spirifer zone, but particularly in the Strophonella faunule, where the species is abundant.

Plesiotypes. — Nos. 8103 to 8114, University of Michigan.

STROPHONELLA HYBRIDA form QUADRATA, nov.

(Plate XXIII, Figs. 16–18)

Description. — Shell of medium size, with greatest width along or slightly anterior to the hinge-line. Dimensions of the holotype: length of pedicle valve, 25.2 mm.; length of brachial valve, 23.2 mm.; width at hinge-line, 29 mm.; width 15 mm. anterior to the hinge-line, 27.8 mm.; thickness, 11.8 mm.

This form differs from *S. hybrida* in its smaller size, greater length in proportion to width, almost straight lateral margins, flattened umbonal region in the brachial valve, and very slightly curved anterior margin. The plications in some specimens are slightly coarser than those of the typical *hybrida* and in others quite as fine; the interior markings of the brachial valve are those of a mature but flattened and elongate *hybrida*. The interior of the pedicle valve is not known. Externally it closely resembles *S. reversa lata* of this volume, but is heavier, longer, flatter, shallower, and more finely costate than that form.

Occurrence. — Probably throughout the Spirifer zone. The sets in our collections contain material from the middle and upper beds of that member at Rockford, the hills northwestward from Rockford, and Bird Hill.

Holotype. — No. 8121; *Allotype.* — No. 8122; *Paratype.* — No. 8123, University of Michigan.

STROPHONELLA HYBRIDA PONDEROSA, n. var.

(Plate XXIV, Figs. 11–15)

Strophonella reversa Hall and Clarke, Pal. N. Y., vol. 8, pt. 1, pl. 12, fig. 19 (no other figures of *S. reversa*). 1892.

Description. — Shell large, semielliptical, wider than long; the greatest width being along the hinge-line. Dimensions of the holotype, which may be considered a typical adult of the variety: length of pedicle valve, 29.2 mm.; length of brachial valve, 29.7 mm.; width at hinge-line, 33.8 mm.; width 14 mm. anterior to the hinge-line, 33.2 mm.; thickness, 11.5 mm.

Pedicle valve slightly convex in the umbonal region, with low, obtuse beak, and broad, striate cardinal area. The anterior portion is shallowly concave, with or without mesial ridge; cardinal extremities flattened. Interior of the valve very coarsely marked, and the hinge-teeth heavy. The diductor scars are broadly flabellate, with the distinction between the anterior and posterior muscle bases indistinct. The adductor scar is deep, the cardinal process heavy and irregular, and the sockets deep. Vascular area covered by large genital pustules; extra-vascular ridge thick and high, crossed by numerous deep vascular channels, and in most specimens bisected by a deeper sinus which originates at the termination of the diductor scars.

Brachial valve somewhat flattened at the umbo, depressed at the cardinal extremities, and convex over the remaining portion. In some specimens the extremities show wrinkling similar to that seen in *S. hybrida,* but much coarser. The interior is much like that of *hybrida,* but far heavier. The cardinal processes are very large and heavy, the adductor scars deep and bounded by heavy elevations, and the vascular area divided into two deep, subovate lobes by a mesial elevation or septum. The extra-vascular ridge is high, and the vascular channels are very deep.

Remarks. — There is little difficulty in distinguishing typical specimens of this variety from equally typical specimens of the species *hybrida.* The greater length, thickness and coarseness of internal markings offer excellent criteria. The surface costae of *ponderosa,* although badly worn in most specimens, appear to be heavier than those of *hybrida.* However, as noted under the discussion of *S. hybrida,* there are numerous pedicle valves that may be gerontic *hybrida* specimens, intermediate stages between

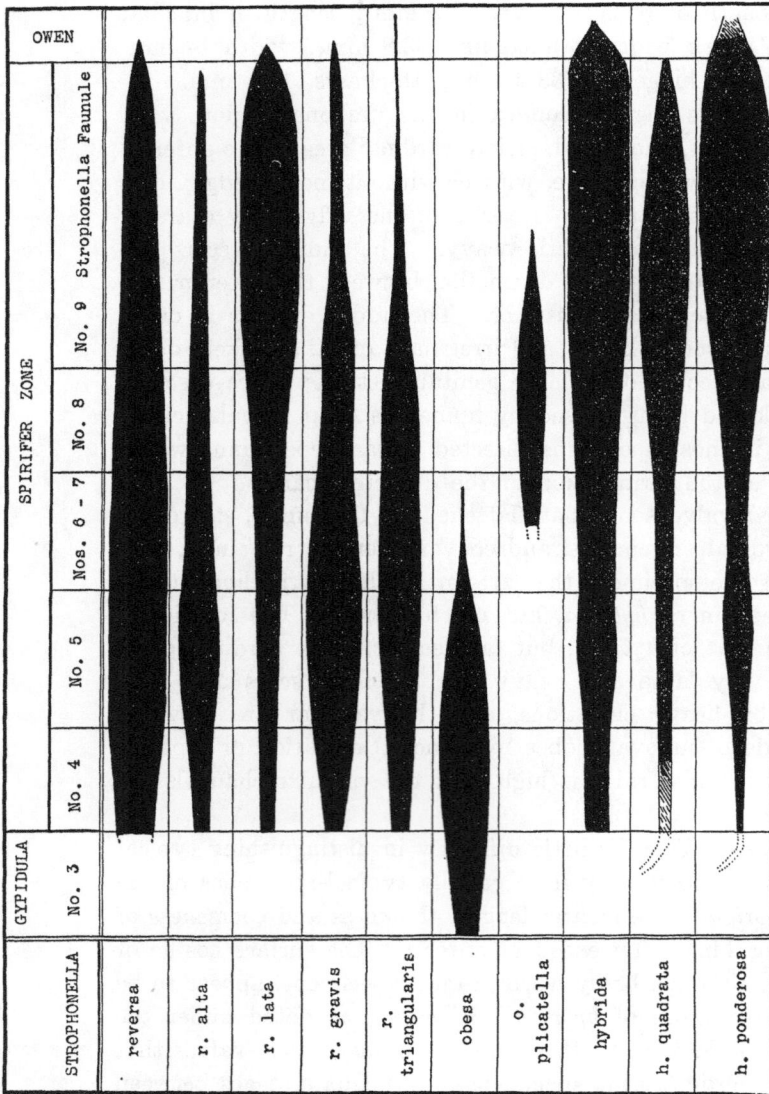

FIG. 7. VERTICAL DISTRIBUTION OF HACKBERRY STROPHONELLAS.

Data based on collections made in 1915–1921. Numbers in section correspond to those of the composite Rockford-Bird Hill section. The diagram shows the preliminary optimum of No. 5 and the final one of No. 9, terminated by rapid extinction in the Owen.

the two forms *hybrida* and *ponderosa*, or, quite conceivably, both. As a matter of convenience, we have referred the longer of these valves to *ponderosa*, and the wider to gerontic *hybrida*.

Occurrence. — If the distinction just made is reliable, the variety *ponderosa* is found in some abundance in those portions of the Spirifer zone which lie above the Pugnoides beds at Rockford and the region westward, and in the Stromatoporella faunule in the neighborhood of Hackberry Grove. The types are from the Strophonella beds at Bird Hill.

Holotype. — No. 8116; *Allotype.* — No. 8117; *Paratypes.* — Nos. 8118, 8119 and 8120, University of Michigan.

Genus SCHUCHERTELLA Girty

SCHUCHERTELLA PRAVA (Hall)
(Plate XX, Figs. 21–28)

Orthis prava Hall, Geol. Ia., vol. 1, pt. 2, p. 490. 1858.
Orthotetes prava Hall and Clarke, Pal. N. Y., vol. 8, pt. 1, p. 255, pl. 11A, fig. 13. 1892.
Orthotetes pravus Schuchert, Bull. U.S. G. S. 87, p. 298. 1897.
Schuchertella parva (misprint for *prava*) Fenton, Am. Journ. Sci., 4th ser., vol. 48, p. 371. 1919.

Description. — Shell small to more than medium size, semi-elliptical in outline, broader than long with the greatest width near the mid-length of shell; cardinal extremities obtusely angular. Dimensions of three typical specimens: length of pedicle valve, 17.3 mm., 16.5 mm. and 10.5 mm.; width, 23 mm., 20.5 mm. and 14 mm.; thickness, 10.2 mm., 5.8 mm. and 4.3 mm. height of cardinal area, 4.8 mm., 3.5 mm. and 3.6 mm.

Pedicle valve moderately convex in large specimens, slightly so in small ones; in most cases it is distorted from adhesion and not uncommonly the entire shell shows this character. Mesial sinus nearly obsolete; cardinal area moderately high, inclined posteriorly without arching. Deltidium strongly convex, marked

by heavy growth lamellae. The muscular impressions correspond to the umbonal elevation and though deep are indistinctly
outlined. There is only a trace of the vascular markings in the
larger specimens, while in the smaller one they are indistinguishable. The cardinal teeth are unsupported by dental plates,
although the inner surface of the valve is thickened along each
side of the deltidium. The inner margin is marked by crenula-
tions laterally and anteriorly; these correspond to the costae.

Brachial valve more convex than the pedicle in the large
specimens. In the smaller ones it is either concave or convex
with the concavity reaching the mid-length of the shell and less
often the anterior margin. The umbo is small and does not
extend beyond the cardinal margin; the mesial area is, in most
cases, slightly flattened from the umbo to the anterior margin;
some specimens show a slight indication of a mesial fold. Cardinal area linear; cardinal process short and rather heavy, bifid,
with a chilidium adjacent to the cardinal margin and occupying
the concavity of the delthyrium. Crural plates prominent;
dental sockets deep.

Surface of both valves marked by subangular or rounded
radiating costae that increase by bifurcation, intercalation and
even trifurcation, from 2 to 3 occupy the space of 1 mm. at the
anterior of mature shells. The costae are crossed by fine concentric markings and by much stronger concentric lines of
growth which are distributed evenly from the mid-length to the
anterior margin.

Remarks. — The species *prava* was described rather indefinitely by Hall; from the description alone identification would
be quite impossible. The specimen figured by Hall and Clarke
is not typical of the species, as may be seen by reference to the
illustrations. However, the few specimens found that do show
the strongly ridged muscular area suffice to prove the identity
of the species.

Occurrence. — *Schuchertella prava* occurs throughout the Spirifer zone of the Hackberry, being especially common in the Strophonella faunule at Bird Hill, and in the corresponding beds at Hackberry Grove.

Plesiotypes. — Nos. 7946 to 7950, University of Michigan.

GENUS PRODUCTELLA HALL

PRODUCTELLA WALCOTTI, nom. nov.

(Plate XXVI, Figs. 1–7)

Productus dissimilis Hall, Geol. Ia., vol. 1, pt. 2, p. 497, pl. 3, fig. 7. 1858.
Productus (Productella) hallanus Walcott, Mon. U.S. G. S., vol. 8, p. 130 (part), not pl. 13, fig. 17. 1884.

Description. — Shell under medium size, wider than long or longer than wide, with the greatest width near the mid-length of the shell; cardinal extremities angular. Dimensions of three typical specimens: length of pedicle valve, 12.5 mm., 15.3 mm. and 15.5 mm.; length of brachial valve, 10.8 mm., 13.7 mm. and 12.6 mm.; greatest width, 14.9 mm., 14.8 mm. and 14.1 mm.; thickness, 5.8 mm., 6.4 mm. and 7.2 mm.; length of hinge-line, 13.9 mm., 12.7 mm. and 11.2 mm.

Pedicle valve highly convex; cardinal area linear; beak strongly incurved; umbonal region prominent and elevated above the hinge-line. Surface curves abruptly to cardinal, and less so to the lateral and anterior margins; postero-lateral surfaces deflected to form small flattened auriculations.

Brachial valve slightly concave in young specimens and moderately concave in mature ones; postero-lateral portions flattened to correspond to the auriculations of the opposite valve. From the base of the very large bifid cardinal process arises a pair of indistinct, low, broadly diverging ridges that extend about two-thirds of the distance to the postero-lateral margins and around the shell less than 1 mm. from the margin; the abrupt posterior slopes of the ridges form rudimentary dental sockets. Median septum originates about 1.5 mm. from the junction of the diverging ridges and extends beyond the middle of the valve; it is very narrow, low posteriorly and high anteriorly. On the marginal slope, anterior to the medium septum, and, in most cases, on either side of the median septum, appear spine-like ridges.

Surface of pedicle valve marked by fine, rounded, nodose, radiating striae, about 5 to the mm., and a few irregular concentric growth lines that, in some cases, become wrinkles on the cardinal lateral margins. Bases of tubular spines scattered irregularly over the shell with their greatest development on the lateral slopes and hinge-line. One spine measures 1.3 mm. Surface of brachial valve marked by regular concentric imbricating lines with only a trace of radiating striae in some specimens.

Remarks. — Since the publication of Walcott's monograph on the Eureka fauna most paleontologists have considered the species called *dissimilis* by Hall to be equivalent to the Eureka species called *hallana* by Walcott. We have examined, however, authentic specimens of *hallana* from the Eureka district, as well as specimens from the Great Slave Lake exposures. The differences between them are distinct. Of the two, *P. hallana* is the larger and somewhat heavier shell. Its surface is slightly less convex in the umbonal region and more so in the anterior portion than is that of *P. walcotti*. The concentric growth lines of the pedicle valve are coarser and, in most cases, the spine bases are heavier than those in *P. walcotti*.

Occurrence. — The species *walcotti* seems to be restricted to the Hackberry stage. It appears, so far as can be determined, in the Gypidula faunule, and reaches its maximum abundance in the Strophonella zone.

Plesiotypes. — Nos. 7951 to 7955, University of Michigan.

PRODUCTELLA HALLANA Walcott

(Plate XXVI, Fig. 8)

Productus (Productella) hallanus Walcott, Mon. U.S. G. S., vol. 8, p. 130, pl. 13, fig. 17. 1884.

In order to show the differences between *P. walcotti* and *P. hallana*, figures of the latter are introduced. Though the specimen is not a cotype, it was identified by Dr. Bassler and is from the type locality.

GENUS GYPIDULA HALL

GYPIDULA CORNUTA, n. sp.

(Plate XXV, Figs. 26–31)

Gypidula comis munda C. L. Fenton, Am. Journ. Sci., 4th ser., vol. 48, p. 372. 1919.
Gypidula comis, n. var. C. L. Fenton, loc. cit.

Description. — Shell of medium size or less, wider than long in young specimens and longer than wide in old ones. Dimensions of three specimens, the second of which is the holotype: length of pedicle valve, 16.7 mm., 21.8 mm. and 23.6 mm.; length of brachial valve, 15.5 mm., 18.9 mm. and 19.5 mm.; width, 19.2 mm., 21.8 mm. and 23.3 mm.; thickness, 9.2 mm., 14.8 mm. and 16.2 mm.

Pedicle valve highly convex; beak large, prominent, sharply pointed, strongly incurved. Cardinal area broadly triangular, strongly arched; pedicle opening seen only in young specimens, triangular and about as high as wide. Umbonal region high; postero-lateral slopes slightly flattened and concave; lateral slopes descend abruptly from the mesial portion, anterior margin sinuate. Mesial fold originates about 10 mm. from the beak and is low and broad, or narrow and prominent; scarcely distinguishable in small specimens since they are quite flat.   On the fold are 2 or 3 low rounded plications separated by moderately broad shallow furrows.   Lateral slopes of large specimens smooth or marked by 2 or more very low rounded plications.

Brachial valve moderately convex in umbonal region; beak pointed, slightly incurved in small specimens but beneath the beak of opposite valve in mature ones.   Cardinal area very low and slightly arched.   Umbonal area moderately convex; posterolateral slopes flattened and slightly concave.   Mesial sinus originates from 5 to 8 mm. anterior to the beak, becomes broad but not deep; only slightly developed in the young individuals. Sinus of the mature specimens bears 1 or 2 low rounded plications;   furrows broad, shallow and rounded.   Lateral slopes smooth, or as in the case of the allotype 2 or more low rounded plications.

Surface of both valves marked by fine concentric lines and heavier growth wrinkles.

Remarks. — This species has usually been referred to Owen's species *comis.*   That form, however, is distinctly a Cedar Valley species, probably not even reaching the upper members of that formation.

Occurrence. — *Gypidula cornuta* appears in the Hackberry in the Gypidula faunule, being common at Rockford and not rare at other localities.   Apparently it was gregarious in habits, for the Gypidula shales contain considerable colonies of the species; in one case 11 were found in a surface of not more than 40 square inches.   In the Spirifer zone the species is much less common, while in the uppermost beds, at Bird Hill, Hackberry Grove and other localities, it is rare.

Holotype. — No. 26063;   *Allotype.* — No. 26062, Walker Museum;   *Paratypes.* — No. 7959, University of Michigan.

GYPIDULA CORNUTA PARVA, n. var.

(Plate XXV, Figs. 32–35)

Description. — In general shape this form is very similar to the mature specimens of *Gypidula cornuta*, but in size compares with the young individuals. Dimensions of holotype: length of pedicle valve, 16.4 mm.; length of brachial valve, 11.3 mm.; width, 13.8 mm.; thickness, 10 mm.

Gypidula cornuta parva has 2 plications on the mesial fold that are higher and more pronounced than those in *cornuta;* a distinct furrow at either base of the fold that corresponds to definite short rounded plications on the brachial valve. There is only a trace of this furrow in *G. cornuta* and the corresponding plications of the brachial valve discernible in only the allotype.

Occurrence. — Upper beds of the Spirifer zone; holotype from the Strophonella faunule at Bird Hill.

Holotype. — No. 26064, Walker Museum.

GENUS CAMAROTOECHIA HALL & CLARKE

CAMAROTOECHIA SAXATILIS (Hall)

(Plate XXV, Figs. 13–17)

Rhynchonella (Stenocisma) Contracta var. *saxatilis* Hall, Pal. N. Y., vol. 4, p. 417, pl. 54A, figs. 44–51. 1867.

Description. — Shell small, wider than long. Dimensions of the specimen illustrated and one other: length of pedicle valve, 8.9 mm. and 9.5 mm.; length of brachial valve, 8.8 mm. and 8.7 mm.; width, 12.5 mm. and 10.5 mm.; thickness, 7.6 and 7.5 mm.

Pedicle valve convex in the umbo, with sharp beak that may or may not extend well beyond that of the brachial valve. Lateral slopes depressed to concave; sinus broad and deep, but not sharply defined. Plications on the lateral slopes number 5 to 8, with 6 as the usual number; those on the sinus 3 to 5. All plications, and particularly those on the sinus, are strong and angular. Brachial valve rounded in the umbo, with the beak incurved; lateral and postero-lateral slopes abruptly convex. Fold apparent only in the anterior half of the valve, where it is rather flattened, though high. Plications correspond to those of the opposite valve.

Remarks. — This species was compared by Hall to the eastern species *contracta* and *eximia*. From the former it differs in size, in lesser strength of plications, and the more abrupt lingual extension of the sinus. It is distinguished from *eximia* by its much smaller size and, on the average, lesser number of plications of fold and sinus. Clearly this is another of the many Hackberry species that, though closely related to species of the eastern provinces, is still sufficiently distinct to warrant separation.

Distribution. — Throughout the Spirifer zone, but particularly in the portions above the Leptostrophia faunule. It may occur in the Gypidula faunule as well.

Plesiotype. — No. 8132, University of Michigan.

Genus LEIORHYNCHUS Hall

LEIORHYNCHUS IRIS Hall

(Plate XXVI, Fig. 27)

Leiorhynchus iris Hall, Pal. N. Y., vol. 4, p. 360, pl. 56, figs. 41–43. 1867.

Description. — Shell small, subangularly ovoid. Dimensions of the plesiotype: length of pedicle valve, 16.2 mm.; length of brachial valve, 15.3 mm.; thickness, 15 mm.

Pedicle valve with prominent umbo and sharply incurved beak. Mesial sinus broad and deep, with strong upwardly directed lingual extension. Antero-lateral slopes bear 3 strong but rounded plications; the sinus, 2 to 4 very weak ones. Postero-lateral slopes compressed. Brachial valve highly convex in the umbo, with beak incurved into the pedicle opening of the opposite valve. Fold appears in the anterior half of the shell, and is high. Plications correspond to those of the opposite valve. Surface marked by concentric lines of growth.

Remarks. — The locality and horizon for the type were given as "calcareous shales of the age of the Hamilton or Chemung formations, near Rockford, Indiana." For 'Indiana' should be read 'Iowa.' The specimen in hand is considerably larger than the one figured by Hall, and less perfect, but the species seems to be the same.

Occurrence. — *L. iris* appears in the Gypidula faunule and extends into the Spirifer zone. It is uncommon.

Plesiotype. — No. 8131, University of Michigan.

Genus PUGNOIDES Weller

PUGNOIDES CALVINI, nom. nov.

(Plate XXV, Figs. 1–8)

Rhynchonella alta Calvin, paper read before the Ia. Acad. Sci., and photographic plate distributed but not published. 1876.
Pugnax altus Hall and Clarke, Pal. N. Y., vol. 8, pt. 2, pl. 60, figs. 4, 5 (*not* figs. 1–3). 1894.
Pugnoides altus Thomas and Stainbrook, Proc. Ia. Acad. Sci., vol. 29, p. 95, pl. 1, figs. 1–16. 1924.

Description. — Shell below medium size, broadly ovate-subtriangular in outline, wider than long; greatest width near or posterior to the mid-length. Postero-lateral margins nearly straight and meeting in an obtuse angle; antero-lateral margins rounded; anterior margin sinuate. Dimensions of three nearly perfect specimens: length of pedicle valve, 10.3 mm., 8.6 mm.

and 11.8 mm.; length of brachial valve, 9.8 mm., 8.4 mm. and 12.9 mm.; width, 13.2 mm., 11 mm. and 17.1 mm.; thickness, 8 mm., 12 mm. and 15.6 mm.; length of hinge-line, 4.9 mm., 5.3 mm. and 8.4 mm.; width of mesial sinus of pedicle valve in front, 10.7 mm., 8.7 mm. and 12.3 mm.

Pedicle valve gently convex in the umbonal region, flattened toward the antero-lateral margins, very strongly arched from beak to front along the median line, the curve being semicircular in the high specimens but semiovate in the low ones. Along the postero-lateral margins toward the beak, the edge of the valve is abruptly inflected to form a sort of false cardinal area. Beak obtusely pointed and moderately incurved; pedicle opening round, apical; mesial sinus originates about 4 mm. from the beak, becomes broad and moderately deep anteriorly and produces a broadly rounded lingual extension, the surface of which is directed upward, in older specimens forming a right angle or even more than a right angle with the plane of the valve. In some specimens this reversal is so great as to amount to from 110 to 120 degrees. Plications obsolete upon the posterior portion of the valve; strongly developed beyond the middle of the large specimens, but only slightly developed in the young ones. Sinus bears from 2 to 4 high, angular plications separated by broad but angular furrows. Plications of lateral slopes number from 1 to 4 — most commonly 2 or 3 — and are higher, narrower and more acutely angular than those of the sinus.

Brachial valve very highly convex transversely and very slightly convex from beak to anterior margins along the median line; mesial fold appears at the beak in some specimens and about 5 mm. anterior in others. Beak obtusely angular and incurved beneath the beak of the opposite valve. From 2 to 3 rather short but heavy angular plications upon the anterior portion of fold, usually a short and fainter one on each lateral slope; occasionally there are 4 heavy ones on the fold with none on the lateral slopes; the postero-lateral margins bear from 1 to 3 plications indistinct or heavy and varying in size.

Surface of both valves marked by very fine radiating striae. Fine concentric lines of growth observed on one specimen; these were near the anterior margin and extended down on the slope of the sinus.

Remarks. — This species has commonly gone under the name

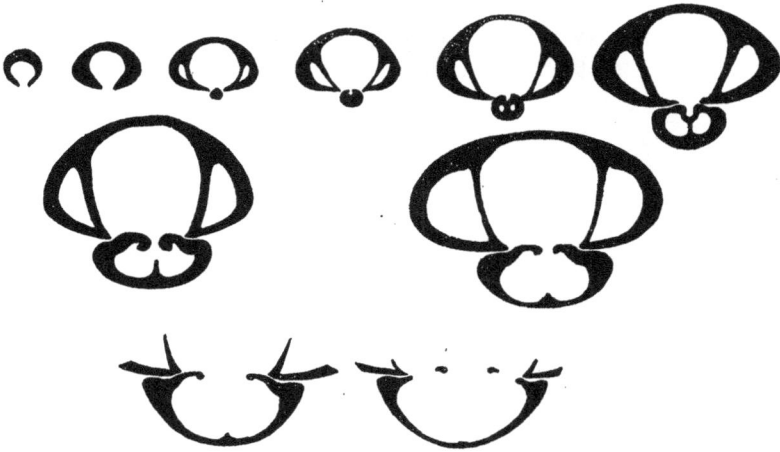

Fɪɢ. 8. Sections of *Pugnoides calvini*, Enlarged

Pugnax altus (Calvin). Sections (see Fig. 8), however, show it to belong to the genus *Pugnoides* Weller, while an examination of such illustrations and descriptions as have been published indicates that the name *altus* must be dropped as preoccupied. In the first place, Calvin did not actually publish his species; he read a description at a meeting of the Iowa Academy, and distributed a named photographic plate. The form illustrated by Williams as *Rhynchonella pugnax alta* (*Bull. G. S. A.*, vol. 1, Pl. 12, Figs. 5–7) differs specifically from Calvin's original *altus*, as is shown by Thomas and Stainbrook in their paper of 1924.

But, as these authors themselves point out, Calvin himself was not consistent in his use of *altus*, confusing under it both of the species here called *calvini* and *solon*.

Hall and Clarke also illustrate the two distinct forms as *Pugnax altus*, in Part 2 of Volume 8 of the *Paleontology*, and it must be admitted that their publication is quite as good as that of Williams and better than the mere mention of names to be found in the writings of Calvin. Thus it is plain that (1) the "species" *altus* was neither defined nor illustrated in print by its author, and (2) that such early mention as was made of it either placed it as synonymous with the European *P. pugnus* (Martin), or confused it with another species, or both. Under such circumstances it seems impossible to accept the name at all, even on the basis of the excellent description and figures by Thomas and Stainbrook. Whatever weight there is in early publication is on the side of the Solon form, and even there it is inadequate. Accordingly we abandon *P. altus* Calvin as without publication by the author, and without recognizable publication by others. *P. altus* Thomas and Stainbrook becomes a homonym of *P. altus* of other writers, and a synonym-homonym of the *altus* of Hall and Clarke, undefined. *P. solon* Thomas and Stainbrook, on the other hand, is a valid name, since the publications of Williams and Hall and Clarke have failed to establish *altus* for the Solon species. To *Pugnoides altus* as used by Thomas and Stainbrook we give the name *Pugnoides calvini*.

Occurrence. — *Pugnoides calvini* is distinctly a Hackberry species, and neither the collections of C. L. Webster nor those of the authors contain examples from the State Quarry beds. In the Hackberry the species appears to be restricted to the Spirifer zone, being particularly abundant in the Pugnoides and Strophonella faunules at Bird Hill.

Holotype. — No. 7961; *Paratypes.* — Nos. 7962 to 7964, University of Michigan.

PUGNOIDES SOLON Thomas & Stainbrook

(Plate XXV, Figs. 9–12)

Rhynchonella pugnus var. *altus* Williams, Bull. G. S. A., vol. 1, pl. 12, figs. 5–7. 1890.

Pugnax altus Hall and Clarke, Pal. N. Y., vol. 8, pt. 2, pl. 60, figs. 1–3 (*not* 4–5). 1893.

Pugnoides solon Thomas and Stainbrook, Science, New Ser., vol. 54, p. 308. 1921.

Pugnoides solon Thomas and Stainbrook, Proc. Ia. Acad. Sci., vol. 29, p. 97, pl. 1, figs. 17–32. 1924.

Remarks. — Inasmuch as this is not a Hackberry species, a description of it will not be given here. The drawings are introduced for comparison with those of *P. calvini.* It will be sufficient to give the following measurements, taken from two typical specimens collected at Solon, Ia.: length, 11.6 mm. and 12.8 mm.; width, 15 mm. and 17.4 mm.; thickness, 11.7 mm. and 13.2 mm. From these it will be seen that, in proportion, *P. altus* differs from *P. calvini* in greater width, lesser thickness, more abrupt, flat-bottomed sinus, smoother umbonal region, flatter, lower, and more flattened fold, and greater general compactness of proportion.

Occurrence. — Limestones of the State Quarry formation, particularly at Solon, Johnson County, Iowa; upper Devonian, Missouri.

Plesiotypes. — Nos. 7965 and 7966, University of Michigan.

GENUS CRANAENELLA NOV.

Genotype: TEREBRATULA NAVICELLA Hall

Description. — Shell terebratuliform, small to large, and profusely punctate. In general appearance the members of this genus are nearly identical with *Cranaena* of Hall and Clarke. From that genus, however, *Cranaenella* differs in that the shells possess a heavy calcareous pedicle tube, to which the name *false syrinx* is here applied. This tube originates at the apex of the beak, and continues forward for as much as 3 mm. In *C. calvini* the tube is united with the shell wall by a secondary calcareous deposit, but is distinct even then. This tube is the diagnostic character of the genus, and is not found in *Cranaena iowaensis*, the genotype of *Cranaena*. The internal characters of the genus are shown in the accompanying sections (Fig. 9).

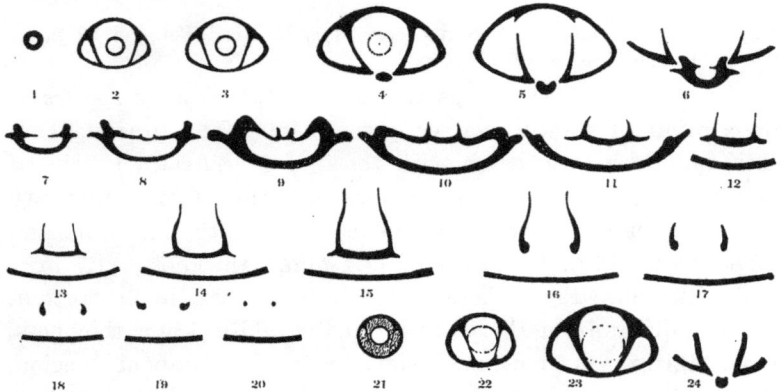

FIG. 9.   Sections of *Cranaenella*

Nos. 1–20 represent sections of the genotype, *C. navicella;* 21–24 show sections from the beak of *C. calvini.* The sections of the latter are not given in full because the main part of the shell presents a close resemblance to the genotype.

CRANAENELLA NAVICELLA (Hall)

(Plate XXV, Figs. 18–22)

Terebratula navicella Hall, Pal. N. Y., vol. 4, p. 391, pl. 60, figs. 38–44. 1867.

Centronella (?) navicella Hall and Clarke, Pal. N. Y., vol. 8, pt. 2, p. 79, figs. 40–42. 1895.

Centronella navicella C. L. Fenton, Am. Journ. Sci., 4th ser., vol. 48, 372. 1919.

Description. — Shell small, subovate, longer than wide, with greatest width anterior to the mid-length. Anterior margin semicircular or semielliptical; lateral margins straight or moder-

ately convex. Dimensions of the three figured specimens:
length of pedicle valve, 11.8 mm., 12.5 mm. and 14.3 mm.;
length of brachial valve, 10.5 mm., 11 mm. and 12.8 mm.;
width, 9.5 mm., 10.8 mm. and 11.9 mm.; thickness, 5.2 mm.,
5.1 mm. and 5.4 mm. All of these are large specimens.

Pedicle valve longitudinally subcarinate, with prominent,
strongly convex umbonal region, slightly convex lateral slopes,
and flattened or depressed mesial region anteriorly. This depression, with an anterior, sinus-like projection of the valve, is
developed only in mature specimens. Beak heavy, prominent,
arched toward the brachial valve, but not incurved; foramen
large, round, apical. Postero-lateral margins incurved to form
a false cardinal area of slight extent.

Brachial valve less convex than the pedicle, with the greatest
depth at or posterior to the mid-length. Umbonal region moderately convex; beak acutely pointed. Antero-lateral slopes slightly
convex or flattened; mesial region broadly elevated, in adult and
gerontic specimens forming a distinct fold which corresponds to
the depressions in the opposite valve.

Surface of valves coarsely punctate, the punctae being so
large and so numerous as to give the shell a strikingly reticulate
appearance when examined under the magnifier. Both valves
marked also by numerous heavy, concentric growth lines and
wrinkles. The internal characters of the shell are discussed
fully in the description of the genus.

Remarks. — This species is the commoner of the two referred
to this genus and found in the Hackberry. In its younger
stages it cannot be separated from the form *calvini* without
grinding the pedicle beak; in the adult ones it is generally
distinguished by its smaller size, more cuneiform shape, heavier
beak, and strong growth lines. These keys, however, are not
always reliable when the specimens are crushed, or swollen by
frost; the one truly diagnostic character is the development and

attachment of the tube in the pedicle valve — the false syrinx. In *navicella* this is fused with the walls of the valve, or with the dental lamellae, only in the very young forms; it reaches a length of as much as 2 mm. In *calvini* the tube, while easily distinguishable, is always attached, and has not been observed to attain a length of more than 1.3 mm. This species is one of the few Hackberry forms to show color, all specimens being of brown, and olive-brown color, with purplish-brown mottlings.

Occurrence. — Found throughout the Spirifer zone of the Hackberry, although uncommon in the lower portions. In the upper beds, particularly those above the Floydia faunule at Rockford and Bird Hill, it is abundant, far outnumbering its related species.

Plesiotypes. — Nos. 7967 to 7969, University of Michigan.

CRANAENELLA CALVINI (Hall & Whitfield)

(Plate XXV, Figs. 23–25)

Cryptonella eudora Hall and Whitfield, 23d Ann. Rep. N. Y. State Cab. Nat. Hist., p. 225. 1873.
Cryptonella calvini Hall and Whitfield, op. cit., p. 239.
(?) *Dielasma calvini* Hall and Clarke, Pal. N. Y., vol. 8, pt. 2, p. 296 (*not* pl. 80, figs. 20–22). 1893.
Cranaena calvini C. L. Fenton, Am. Journ. Sci., 4th ser., vol. 48, p. 372. 1919.

Description. — Shell broadly subovate, longer than wide, with greatest width slightly anterior to the mid-length. Dimensions of a typical adult shell: length of pedicle valve, 27.8 mm.; length of brachial valve, 24.6 mm.; width, 15.7 mm.; thickness, 5.4 mm. Immature specimens maintain about the same proportions.

Pedicle valve moderately convex, with flattened lateral slopes and lessened convexity anteriorly. Beak large, heavy, somewhat incurved in large specimens; foramen apical, round and large. False cardinal area large, arched and constricted at the beak. Mesial portion somewhat more convex than the slopes, projected anteriorly, but without a defined fold; the same is true of the brachial valve, which is shallow, with broad, flattened slopes. Surface marked by reticulate punctae, as in the species *navicella;* growth lines numerous but fine.

Like its smaller relative, *C. calvini* ordinarily preserves much of its coloration. In general this is brownish to olive-brown, with touches of purple and chocolate. In one specimen the ground color is olive-brown, over which are dots and longitudinal splotches of purple. Another, in the collection of C. H. Belanski, shows much the same pattern, with the purple deeper than in our specimen. So far as can be told, *calvini* differs from *navicella* in having more olive in the ground color, and less purple, the latter being restricted to the markings.

Remarks. — In general, *C. calvini* differs from the smaller species in lesser width, lack of fold and sinus, different coloration, and internal characters, which are discussed in the diagnosis of the genus.

Occurrence. — Throughout the Spirifer zone, but particularly in the middle and upper portions. Uncertain in the Owen; probably lacking in the Gypidula faunule. Identifications of the species from other formations is uncertain. It seems probable that the specimens from the Cedar Valley, illustrated by Hall and Clarke as *Dielasma calvini,* belong to the species *iowaensis,* type of *Cranaena.* Whiteaves's identification of Canadian forms was based on too fragmentary material to afford reliable data.

Plesiotypes. — Nos. 8128, University of Michigan, and 25580, Walker Museum.

GENUS ATRYPA DALMAN

ATRYPA DEVONIANA Webster

(Plate XXVI, Figs. 16–24)

Atrypa reticularis Hall, Pal. N. Y., vol. 4, p. 321. 1867.
Atrypa reticularis Calvin, Ia. Geol. Surv., vol. 7, p. 167. 1897.
Atrypa reticularis C. L. Fenton, Am. Mid. Nat., vol. 5, p. 216. 1918.
Atrypa reticularis C. L. Fenton, Am. Journ. Sci., 4th ser., vol. 48, p. 372.
  1919.
Atrypa devoniana Webster, Am. Mid. Nat., vol. 7, p. 19 (*not* pl. 8, figs. 9–11).
  1921.

Description. — Adult shells of more than medium size, wider than long, with greatest width anterior to the hinge-line in all but strongly alate specimens. Dimensions of three typical specimens: length, 20.3 mm., 27.8 mm. and 31.1 mm.; width, 22.7 mm., 31.4 mm. and 35.6 mm.; thickness, 13 mm., 17.4 mm. and 19.0 mm. Of the lengths, about .5 mm., on the average, is consumed by the projection of the pedicle beak, although in large, old specimens the pedicle valve commonly is shorter than the brachial — a condition exhibited in the third of the shells mentioned above.

Pedicle valve much less convex than the brachial; greatest convexity in the umbonal region. Lateral slopes flattened to pronouncedly concave; mesial depression shallow and ill-defined in young specimens, and lacking in the very young; in adults and senescent shells it is deep and fairly well defined. Beak prominent, incurved; foramen small, round and apical. Interiors of pedicle valves commonly poorly preserved. The diductor scars are broad and poorly defined, while those of the adductors are almost indistinguishable. The teeth are broad and high and the sockets deep. The deltidial plates close but a small portion of the delthyrium, leaving the rest to be occupied by the beak of the brachial valve.

Brachial valve but little more convex than the pedicle in small specimens; with increase of size the convexity becomes greater, until in old shells the brachial valve is from four to six times as deep as the pedicle. Convexity greatest in the umbonal region; lateral and postero-lateral slopes commonly extended, flattened, and even concave in senescent individuals. Mesial portion of the valve marked by a poorly bounded, round fold, which becomes higher and more distinct in aged specimens. Interior of the brachial valve shown in the plate.

Surface of both valves marked by fine, radiating, dichotomizing plications, which become finer as they divide. Near the beak, 10 to 12 of these plications occupy the space of 1 cm.; nearer the margin, 14 to 16 occupy the same space. The plications are crossed by numerous concentric lamellae which increase in number toward the margin. On the umbonal and medial regions, these lamellae are from 2 to 4 mm. apart, while near the margin they are so closely placed that they overlap. On the alate shell extensions, however, they again become less numerous, single lamellae 5 mm. in width not being uncommon. Unlike the lamellae, however, in *Atrypas* of the *aspera-hystrix-rockfordensis* type, these lamellae do not form spines, the extensions being mere open folds.

Atrypa devoniana, in common with other species, exhibits a strong tendency toward alateness in the senescent shells. This stage is marked by broad extensions on the postero-lateral slopes, and an increased development of the fold and sinus. Such a development is shown in the largest of the plesiotypes, although in this specimen the extensions are limited. The full width of the extensions is seldom preserved; they evidently are very fragile, and are destroyed by weathering. As will be noted farther on in this volume, alateness is not a characteristic of all specimens of senescent *devoniana*, but it is by far the most common type of old-age development.

Remarks. — *Atrypa devoniana* is by far the most common brachiopod of the Hackberry stage. By most authors it has been identified with *Atrypa reticularis* (Linne), a species which, although named more than a century ago, remains without satisfactory definition. Mr. Webster described the Hackberry form accurately, but through an unfortunate mistake at the printer's, his illustrations were confused with some others, and were not published. The species is, however, sufficiently distinct to make possible off-hand identification.

Occurrence. — *Atrypa devoniana* is common to abundant throughout the Cerro Gordo substage, and probably extends throughout most of the Owen as well. Whether or not it extends below the Gypidula faunule is uncertain, inasmuch as all specimens from that member in the authors' collections are identifiable as *A. hackberryensis*. However, casts from the lowest beds of the Hackberry at Mason City appear to belong to Webster's species.

Cotypes and *Paratypes.* — Collection of C. L. Webster; *Plesiotypes.* — Nos. 7976 to 7981, University of Michigan.

ATRYPA DEVONIANA, form ALTA C. L. Fenton

(Plate XXVI, Figs. 25–26)

Atrypa reticularis alta C. L. Fenton, Am. Mid. Nat., vol. 5, p. 216, pl. 6, figs.1–2. 1918.

Description. — Shell of medium or less than medium size; greatest width about 5 mm. anterior to the hinge-line. Dimensions of the holotype: length, 23.6 mm.; width, 24 mm.; thickness, 19.6 mm.

In character of markings this shell agrees with *A. devoniana*, except that the plications are somewhat finer than in the typical examples of that species. Both beaks abruptly incurved; mesial sinus and fold pronounced, with a crowding of growth lamellae toward the anterior margin. Brachial valve very deep — about 17.5 mm. in the holotype. Along the mesial portion runs the elevated ridge that constitutes the fold, this ridge being the most distinctive feature of the shell.

Remarks. — In the description of the form *alta* it was considered to be a variety. Similar development, however, is found in a number of other *Atrypas* that cannot possibly be considered to be of the species *devoniana*. The conclusion, therefore, is that the type *alta* represents a form of old age development common to a number of species, though particularly noticeable in *A. devoniana*, but that this type cannot be considered a true variety. It is of particular interest to note that in this one species, at least, there is more than one sort of senescent growth.

Occurrence. — Spirifer zone of the Hackberry.[17]

Holotype. — No. 26065, Walker Museum.

ATRYPA HACKBERRYENSIS, n. sp.

(Plate XXVI, Figs. 12–15)

Atrypa hackberryensis Webster, MS. and labels.
Atrypa reticularis hackberryensis C. L. Fenton, Am. Journ. Sci., 4th ser., vol. 48, p. 372. 1919.[18]

Description. — Shell of medium size or less, wider than long, with greatest width anterior to the hinge-line. Dimensions of two specimens, the holotype and allotype: length, 18.7 mm. and 23.3 mm.; width, 21.7 mm. and 25.5 mm.; thickness, 9.4 mm. and 11.8 mm. Projection of the pedicle beaks .5 mm. or less.

---

[17] The supposed Cedar Valley specimens of *A. r. alta* prove to belong to a different species, probably *A. lineata* Webster (*Am. Mid. Nat.*, vol. 7, p. 17. 1921).

[18] As in the case of *A. rockfordensis*, I used Webster's manuscript name, since both of us expected his paper to appear before either of mine. In order to prevent unnecessary confusion, the name is here retained. C. L. F.

Pedicle valve moderately convex in the umbonal region; flattened and convex marginally. The mesial portion is depressed anteriorly into a broad, flat-bottomed, sharply defined and angular depression or sinus which is noticeably produced on the anterior margin. Brachial valve about twice as deep as the pedicle; lateral slopes flattened or concave; umbonal region prominent; beak incurved into the delthyrium of the pedicle valve. Postero-lateral slopes concave in mature and old specimens. Mesial elevation or fold broad, low and subangular.

Surface of both valves marked by fine radiating plications which bifurcate within 4 or 5 mm. of the beak. A distinctive feature of this bifurcation is that in most specimens it takes place but once, and in very few cases does it appear more than twice. Furthermore, there is little reduction in size of the plications after bifurcation — a character distinctly in contrast with *A. devoniana*. There are from 14 to 18 plications in the space of 1 cm. Plications are crossed by numerous fine, concentric growth laminae, which fail to show evidence of spines.

Remarks. — This species differs from *A. devoniana* to so great an extent that it deserves more than a varietal rank, as at first supposed by C. L. Fenton. The fine plications, pronounced, subangular fold and sinus, and tendency toward alateness in both young and adult specimens are all distinctive characters. The relatively straight hinge-line, and the lesser proportion of thickness to length and width are other distinctive features.

Occurrence. — Probably throughout the lowest division of the Hackberry. Is particularly abundant in the Gypidula faunule at Rockford.

Holotype. — No. 7973; *Allotype.* — No. 7974; *Paratypes.* — No. 7975, University of Michigan.

ATRYPA PLANOSULCATA Webster

(Plate XXVII, Figs. 13–16)

Atrypa hystrix var. *planosulcata* Webster, Am. Nat., vol. 22, p. 1104. 1888.
Atrypa spinosa (?) C. L. Fenton, Am. Journ. Sci., 4th ser., vol. 48, p. 372. 1919.

Description. — Shell of less than medium size, wider than long, with greatest width about mid-length of the shell. Dimensions of three specimens: length, 18.4 mm., 23.2 mm. and 21.3 mm.; width, 21.4 mm., 26.2 mm. and 22.9 mm.; thickness, 9.2 mm., 13.3 mm. and 15.7 mm. Of the lengths, less than 1 mm. is taken up by the projection of the pedicle beak.

In general form, the valves of this species differ but slightly from those of *A. rockfordensis*. There is a somewhat more pronounced tendency toward gibbosity in *A. planosulcata*, accompanied by a lessening of the postero-lateral concavity, and a development of a broad, poorly defined sinus and fold. Pedicle interiors, however, show some striking differences between the species. The one here considered has much the thicker shell, stronger teeth, and stronger muscular scars. The diductor scars are broadly leaf-shaped, with the adductors represented by a sharp depression. The entire muscular area is sunken deeply and surrounded by a fairly sharp margin. Beyond it, extending well over the valve are the pitted genitalial markings, as well as numerous marginally directed vascular channels.

Surface of both valves covered by rounded plications which number 10 to 13 in shells about 10 mm. in length. One specimen, 14 mm. in length, shows 15–16 plications. At the distance of 8 or 10 mm. from the beak the plications begin to divide. In the specimens measured the plication numbers at the margins are respectively 27, 28 and 29. The shells bear numerous fine, concentric laminae, which number from 20 to 26 on adult specimens. Spines undoubtedly were present during life, but they were small, and are represented in fossils by but a few bases.

Remarks. — The principal points of difference between this and the preceding species lie in the plications. In *A. rockfordensis* these are very coarse and are loosely placed; in *A. multilaminata* they are fine and crowded. Lamellae, too, are more abundant, and much more crowded in the latter species. The internal characters, too, add to the distinction.

Occurrence. — Throughout the Cerro Gordo substage, appearing definitely in the Gypidula beds. Most common in the Floydia beds at Rockford, but uncertain in the Owen.

Plesiotypes. — Nos. 7996 and 7997, University of Michigan.

ATRYPA PLANOSULCATA MINOR, n. var.

(Plate XXVII, Figs. 17–18)

Description. — Shell small, width and length variable in proportion. Greatest width about mid-length of the shell. Dimensions of the cotypes: length, 8.4 mm. and 8.8 mm.; width, 7.8 mm. and 9.6 mm.; thickness, 3.8 mm. and 3.9 mm. Projection of pedicle beak about .5 mm.

In general shape and proportions this species corresponds to a young *A. planosulcata,* the differences appearing in the surface markings. The cotypes bear 18 and 22 radiating plications, while on the smaller specimen the concentric lamellae appear within 1 mm. of the beaks. They are so numerous that the entire surface presents a finely crenulate appearance.

Remarks. — This form differs from *A. planosulcata* in the very numerous and excessively fine plications and lamellae, which appear in unusually small specimens. In view of the fact that even in the youngest stages of both forms this differentiation is clearly marked, it is quite probable that the form *minor* represents a distinct species. Insufficient material, however, prevents such a reference in this report.

Occurrence. — Found in the Gypidula faunule, particularly at Rockford. Is relatively uncommon, and is associated with the more numerous *A. hackberryensis.*

Cotypes. — No. 8001, University of Michigan.

ATRYPA OWENENSIS Webster

(Plate XXVI, Figs. 9–11)

Atrypa owenensis Webster, Am. Mid. Nat., vol. 7, p. 14 (*not* pl. 8, figs. 12–14). 1921.

Description. — Shell small, longer than wide. Two of the cotypes show, as lengths for the pedicle valves, 11.4 mm. and 12.8 mm.; for the widths, 10.5 mm. and 11.5 mm. The third cotype has a length of 11.5 mm. and a thickness of 7.2 mm.

Both valves strongly convex, the brachial less so than the pedicle. Mesial fold and sinus virtually lacking. Surface marked by 22 to 30 coarse, rounded plications, which appear to increase solely by implantation. Concentric growth lines strong and numerous, giving the plications an appearance somewhat like those of *S. planosulcata.*

Remarks. — This is a rare species the description of which has entailed some difficulty. The type specimens are not available, so the writers have had to rely on notes and three photographs made some years ago. There can be little doubt as to the validity of the form; its small size, gibbose valves, lack of fold and sinus, and almost nodose plications, are distinctive. Its origin is uncertain; apparently it is not closely related to any other of the Hackberry species.

Occurrence. — Lower portion of the Owen substage at Owen Grove.

Cotypes. — Collection of C. L. Webster.

ATRYPA ROCKFORDENSIS, n. sp.

(Plate XXVII, Figs. 4–12)

Atrypa asper var. *hystrix* Calvin, Ia. Geol. Surv., vol. 7, p. 167. 1897.
Atrypa rockfordensis C. L. Fenton, Am. Mid. Nat., vol. 5, p. 216. 1918.[19]
Atrypa hystrix C. L. Fenton, Am. Journ. Sci., 4th ser., vol. 48, p. 372. 1919.

Description. — Shell above medium size, wider than long, with greatest width about mid-length of the shell. Dimensions of four specimens, the third of which is the holotype: length, 11.6 mm., 14.8 mm., 23.9 mm. and 26.5 mm.; width, 13 mm., 15.4 mm., 28.9 mm. and 30.2 mm.; thickness, 5.6 mm., 6.8 mm., 13.4 mm. and 15.2 mm. About .6 mm. occupied by the pedicle beak.

Pedicle valve moderately convex in umbonal region but less so marginally, especially in large specimens. In the young, the pedicle valve is more convex than the brachial, while in adults the opposite is true. In large specimens, such as the holotype, there is a decided concavity on the lateral slopes, accompanied by a broad, shallow mesial depression. Beak abruptly incurved over that of the opposite valve; cardinal margin curved pronouncedly in small specimens, but somewhat straightened by the extended lateral slopes in large ones. Lateral margin more curved than the anterior.

The interior of the pedicle valve almost invariably presents an appearance of remarkable smoothness. The teeth are prominent and strong, and the sockets deep. The diductor scars are broad, shallow and poorly defined; the adductor scars are almost indistinguishable. Genitalial markings faint or entirely lacking. Entire inner surface in all specimens marked by ridges and depressions corresponding to the markings of the exterior.

[19] In this reference the species is attributed to C. L. Webster, who had a description ready that, it seemed, would appear before the one by Fenton. Webster's paper, however, has never been published, so the name is used here.

Brachial valve equally convex with, or less so than, the pedicle, in young specimens; in adults it attains a convexity about twice that of the pedicle valve. Chief elevation along the median line of the valve, corresponding to the depression in the opposite valve. Lateral slopes flattened near the margin; slightly concave postero-laterally. Beak abruptly incurved, occupying the delthyrium of the pedicle valve. Surface of both valves marked by broad, rounded plications, which number from 6 to 10 in the posterior half of the shell. About 15 mm. from the beak they begin to bifurcate, so that at the anterior margin they number from 18 to 24. These plications are crossed by strong concentric laminae which increase in number toward the anterior margin. At the junctures of the laminae and plications the shell is produced to form long, hollow spines which are largest toward the umbo and most numerous toward the margin. In most of the small specimens these spines appear only near the margin; in the adults they are found over the anterior two-thirds of the shell in well-preserved specimens. Most commonly, however, the spines are worn away, as well as much of the laminae. An intermediate stage of preservation is shown in the holotype, where the spines are preserved on the lateral slopes, while on the mesial portions the laminae alone remain.

Remarks. — This is the species which most paleontologists, when identifying Hackberry material, have referred to *Atrypa hystrix* Hall. A comparison of specimens with Hall's figures and descriptions, however, shows a number of striking differences. *A. hystrix* has but a few — seemingly less than a dozen — plications, which give off spines only near the margin. The concentric laminae are fewer than in *A. rockfordensis*, and exhibit no tendency to increase anteriorly. The shell, too, appears to be normally smaller than that of the Hackberry species. Furthermore, the restricted range of the true *A. spinosa* [20] would argue

[20] See Hall, *Paleontology of New York*, vol. 4, p. 326.

against the appearance of that species in a distinctly north-western fauna. That Hall himself did not identify the Hackberry form with the one from New York is shown by his statement: "In the higher beds of the series (Devonian) in Northern Central Iowa, which may be of the age of the Portage or Chemung formation of New York, the species identified with *A. reticularis* is more finely costate, while the other form approaches more nearly to the *A. hystrix* of our rocks, having a few coarse plications with spines; these appendages, however, are rarely preserved." [21]

Occurrence. — Abundant throughout the Spirifer zone of the Hackberry stage, but much less common in the Owen. Casts of specimens belonging to this species or the preceding are common in the Schizophoria zone.

Holotype. — No. 7985; *Allotype.* — No. 7986; *Paratypes.* — Nos. 7987 to 7991, University of Michigan.

ATRYPA ROCKFORDENSIS ELONGATA Webster

(Plate XXXII, Figs. 5–7)

Atrypa hystrix elongata Webster, Am. Nat., vol. 22, p. 1104. 1888.

Description. — Shell of medium size, with length greater than width and greatest width anterior to the mid-length of the shell. Dimensions of the holotype: length, 23.1 mm.; width, 19.8 mm.; thickness, 10.7 mm.

Brachial valve equal to the pedicle in length, with about equal convexity. Mesial fold and sinus very indistinct. Lateral slopes of the valves compressed. Brachial valve with 7 and pedicle with 8 broad, low, rounded plications which are crossed by numerous growth laminae. Spines lacking or poorly developed.

[21] *Loc. cit.*

Remarks. — The foregoing description is drawn from the holotype alone. The variety appears to be an aberrant derivation from the typical *A. rockfordensis.*

Occurrence. — Spirifer zone at Hackberry Grove, and probably at other localities.

Holotype. — Collection of C. L. Webster.

ATRYPA SUBHANNIBALENSIS Webster

(Plate XXVII, Figs. 1–3)

Atrypa subhannibalensis C. L. Webster, Am. Mid. Nat., vol. 7, p. 18 (*not* pl. 8, figs. 15–16). 1921.

Description. — Shell small, with greatest width equal to, or greater than, the length. Dimensions of holotype: length of pedicle valve, 17 mm.; length of brachial valve, 16.5 mm.; width, 21.4 mm.; thickness, 13 mm.

Pedicle valve moderately convex, with a strong, rounded median plication extending about two thirds of the distance from the beak to the anterior margin. In the anterior third of the shell there is a shallow, rounded sinus. The brachial valve is more convex than the pedicle, with a broad fold and flattened lateral slopes. Near the beak there are 8 rounded plications, 2 of which are specially strong and extend about two thirds of the way to the anterior margin, where they coalesce for the remainder of their extent. Surface of shell marked by heavy, broad lamellae of growth that quite obscure the plications on the anterior portions of the valves.

Remarks. — The foregoing description was taken from the holotype. Another specimen is somewhat narrower and longer, but shows the same heavy lamellae and obscure plications. Evidently the form is a derivation from *A. rockfordensis,* and is more or less gerontic in nature.

Occurrence. — Upper Spirifer zone at Hackberry Grove, and Owen beds, exact horizon not specified, at Owen Grove.

Types. — Collection of C. L. Webster.

GENUS SPIRIFER SOWERBY

SPIRIFER HUNGERFORDI Hall

(Plate XXIII, Figs. 1–3; Plate XXVII, Figs. 19–24)

Spirifer hungerfordi Hall, Geol. Ia., vol. 1, pt. 2, p. 501, pl. 4, fig. 1 a–k
1858.

Remarks. — Several forms have been included under the name *hungerfordi* of Hall, but as yet no detailed study of the group has been possible. From the work done, however, it appears that several species and varieties must be made. At the present time we can make no addition to the information given in the original description. Figures of some typical specimens are included, as well as of some non-typical forms.

Occurrence. — Common throughout the Spirifer zone, but uncommon to rare in the Owen.

Plesiotypes. — Collection of C. L. and Mildred A. Fenton.

SPIRIFER WHITNEYI Hall

(Plate XXVIII, Figs. 1–9; Plate XXX, Figs. 14–19)

Spirifer whitneyi Hall, Geol. Ia., vol. 1, pt. 2, p. 502, pl. 4, fig. 2. 1858.
Spirifera whitneyi Hall, Pal. N. Y., vol. 4, pp. 245, 417. 1867.
Spirifer whitneyi Hall and Clarke, Pal. N. Y., vol. 8, pt. 2, pp. 24, 37,
pl. 30, figs. 18, 19. 1893.

Description. — Shell of medium or large size, with greatest width along the hinge-line. Dimensions of two large and exceptionally perfect specimens: length of pedicle valve, 17 mm. and 19.4 mm.; length of brachial valve, 16.9 mm. and 18.6 mm.;

width, 30.8 mm. and 36.6 mm.; thickness, 15.7 mm. and 18.7 mm.; height of cardinal area, 5.3 mm. and 4.4 mm.

Pedicle valve moderately convex in the anterior two thirds, with the curvature increasing in the umbonal region. Beak prominent, recurved. Shell curves abruptly to the cardinal margin in the region of the beak, less so laterally, and is flattened near the extremities, which are produced to form sharp, though short mucronations. Mesial sinus distinct at beak; broad, deep and subangular, marked by 4 or 5 plications near the beak, the number increasing to 10 or 15 at the anterior margin. The plication pattern is shown excellently in the several figures. The lateral slopes bear 20 to 30 plications, which are broad, rounded and larger than those of the sinus. Area high, flattened below and arched near the beak, marked by fine vertical striae. Delthyrium broadly triangular. Interior of the valve shows the adductor scar to be almost indistinguishable, and the broad diductor scars faint. The teeth are supported by heavy dental lamellae which extend forward almost half the length of the valve, and diverge at about the same rate as the boundaries of the sinus

Brachial valve less convex than the pedicle, with maximum convexity near the umbo. Shell slopes abruptly to the cardinal margin in the umbonal region, but is depressed laterally. Cardinal area narrow and horizontal; beak sharply defined, bearing 10 to 15 plications. Plications of the lateral slopes, like those of the opposite valve, are simple, broad and rounded, and separated by deep, narrow furrows. Interior of the valve shows the cardinal process to be very broad and heavy, the dental sockets deep, and the socket plates broad and heavy.

Surface of the shell marked by fine, irregular, radiating striae. In the young specimens these striae are more or less smooth and unbroken. In the mature forms, however, they are irregular, more or less dichotomizing, and strongly papillose, the papillae commonly obscuring the striae beneath. In the anterior portions of very large specimens the striae are either completely obscured or obsolete, the papillae alone being discernible.

Remarks. — Neanic specimens of *Spirifer whitneyi* show the hinge-line to be shorter than the greatest width, the cardinal extremities being rounded. In these specimens, also, the striae are especially prominent. The adult development consists of an increase in width, with short production of the hinge-line. In this stage growth lines are uncommon over most of the shell, appearing only near the margin. Old age, or late maturity, is marked by a great increase in thickness and a considerable increase in length of the shell. The lateral margins increase their curvature, or become straightened postero-laterally and highly curved antero-laterally. Growth lines become numerous. The gerontic, or senile, stage is marked by still greater thickening of the shell, accompanied by a pronounced longitudinal compression of both valves, elevation and flattening of the pedicle cardinal area, and a great increase in the frequency of growth lines.

It is probable that authors who have referred to the finer markings of this shell as being strong radiating striae which give the plications a grooved appearance have been misled by examination of slightly worn specimens. In these the outer layer of the shell is partly worn away, so that there do appear to be grooves upon the crests. Since these are lacking, however, on those specimens which retain the papillae in full perfection, they may be regarded as an artificial condition produced by wear only. It is true that in some unworn specimens the plications of the sinus and fold show grooves, which may or may not be due to incipient bifurcation. They are not, however, to be homologized with the fine striae.

Spirifer whitneyi and its varieties are the Hackberry representatives of the group to which belong *S. disjunctus*, *S. verneuili*, *S. archiaci*, and other species both European and Ameri-

can. Several attempts have been made to refer it to *disjunctus*
and *archiaci*, while Hall and Whitfield confused it with *verneuili*.
Comparison of several hundred specimens of *whitneyi* with a
large series of typical *verneuili* has shown that in neanic and
youthful stages both species are much alike in form and in
character of plications, while in adulthood and senility they
differ greatly, *verneuili* being much the larger, heavier, and
more angular. Authentic specimens of the European *disjunctus*
have not been available for comparison. We have not studied
the western forms related or belonging to *S. whitneyi*.

Occurrence. — *Spirifer whitneyi* occurs throughout the Hackberry, but is most abundant in the Spirifer zone. Whether or
not those from the lower beds are true *whitneyi* cannot be determined; they exist only as badly distorted casts that are too
poor to admit accurate identification.

Plesiotypes. — Nos. 8133 to 8142, University of Michigan;
Nos. 26066 and 26067, Walker Museum.

SPIRIFER WHITNEYI GRADATUS C. L. Fenton

(Plate XXVIII, Figs. 10–13)

Spirifer Whitneyi gradatus C. L. Fenton, Am. Mid. Nat., vol. 5, p. 219,
pl. 6, figs. 7–10. 1918.

Description. — Shell above medium size, wider than long,
with greatest width along or slightly anterior to the hinge-line.
Dimensions of the three cotypes: length of pedicle valve,
20 mm., 21 mm. and 24.5 mm.; length of brachial valve, 18
mm., 20.1 mm. and 20.3 mm.; width, 24.3 mm., 27.7 mm. and
31 mm.; thickness, 18.3 mm., 23.8 mm. and 25 mm.; height
of cardinal area, 3.8 mm., 6.2 mm. and 5.1 mm.

Pedicle valve highly convex, the curvature of the median
plane being about 180 degrees. Maximum convexity near the
umbo; beak prominent, pointed and incurved. Lateral slopes

strongly convex, with a slight constriction near the extremities. Sinus broad, deep, and rounded, produced in a lingual extension that reaches from 7 to 14 mm. above the plane of the valve. Cardinal area high, arched, vertically striate; delthyrium broadly, though acutely triangular. The sinus bears 9 to 15 broad, low, subangular plications, which in the umbonal and mesial regions are arranged as in the typical *whitneyi*, while on the lingual extension they bifurcate profusely. This bifurcation originates as a groove on the crest of the plication, and deepens until it completely divides the plication. Lateral slopes bear 18 to 25 coarse plications in the typical specimens, or as many as 30 in the more finely plicate.

Brachial valve strongly convex, but less so than the pedicle. Beak short, heavy, and incurved. Lateral slopes convex except in the immediate region of the extremities, where they are slightly compressed. Mesial fold broad, high, rounded or subangular, with the plications arranged as in the typical *whitneyi*. Plications correspond to those of the opposite valve.

Surface marked by fine striae on the umbos; on the mature portions of the shell these striae give way to radiating rows of papillae, which likewise appear to be arranged more or less radially.[22] On the sinus, which seems to undergo the greatest gerontic aberrations, the papillae appear to be scattered without pattern. Growth lines heavy.

Remarks. — This form appears to belong to the rather common class of old-age brachiopods that assume a mature form during youth, and an old-age one in late youth or early maturity. Thus *gradatus* begins as a normal *whitneyi*, but early in life passes through the typical *whitneyi* stage, and begins to develop a narrow, elongate, tumid form. That the development

---

[22] Such a pattern is illustrated by Hall, *Geol. Ia.*, 1, pt. 2, Pl. 4, Fig. 2c. The concentric striae which he mentions are, however, usually indistinguishable.

is not purely a gerontic one is shown by the existence of youthful specimens possessing the typical *gradatus* shape; that it is not mere individual variation is shown by the abundance of the variety.

Occurrence. — Throughout the Spirifer zone, as well as sparingly in the Owen substage. It is most abundant in the lower beds of the Spirifer zone, both at Rockford and Hackberry, and Owen Grove.

Cotypes. — No. 26024 to 26027, Walker Museum.

SPIRIFER WHITNEYI ROCKFORDENSIS C. L. Fenton

(Plate XXVIII, Figs. 14–16)

Spirifer whitneyi rockfordensis C. L. Fenton, Am. Mid. Nat., vol. 5, p. 219, pl. 6, figs. 3–4. 1918.

Description. — Shell large, wider than long, with greatest width along or anterior to the hinge-line. Dimensions of the paratype: length of pedicle valve, 20 mm.; length of brachial valve, 19.3 mm.; width, 28.5 mm.; thickness, 22.2 mm.; height of cardinal area, 2.8 mm.

Pedicle valve highly convex; beak prominent, sharp and incurved. Cardinal area low and strongly arched; foramen almost as wide as high. Mesial sinus broad, shallow and rounded, produced to a long, rounded, lingual extension. Several millimeters posterior to this extension the sinus becomes reversed, forming a fold, which meets the high rounded fold of the brachial valve.

Remarks. — The form *rockfordensis* is a gerontic development which begins much as does *gradatus*. Though there is a slight gradation between the two, the characteristic of the sinus which reverses to become a fold appears, however, to be sufficiently distinct to merit a name.

Occurrence. — Apparently only in the middle beds of the Spirifer zone, particularly in the vicinity of Rockford.

Holotype. — No. 26069, Walker Museum.

SPIRIFER WHITNEYI SUBSIDUUS, n. var.

(Plate XXX, Figs. 10–13)

Description. — Shell large or medium in size, much wider than long, with the greatest width typically at 2 or 3 mm. anterior to the hinge-line. Dimensions of the holotype and paratype: length of pedicle valve, 16.5 mm. and 19.3 mm.; length of brachial valve, 15.5 mm. and 19.3 mm.; width, 29.2 mm. and 35.2 mm.; thickness, 12.5 mm. and 15.5 mm.; height of cardinal area, 4 mm. and 6.2 mm.

Pedicle valve moderately convex, sloping gently to the anterior and lateral margins and slightly compressed toward the cardinal extremities. Beak prominent, pointed, slightly incurved; area broad, concave below the beak, and vertically striate. Delthyrium as wide as, or wider than, high. Mesial sinus broad, rounded or subangular, or even obtusely angular at bottom, sharply defined. Contains 8 to 14 fine plications, arranged as in *S. whitneyi*, and separated by narrow, angular furrows. Lateral slopes bear 20 to 28 simple, rounded plications, which are separated by narrow, deep, subangular furrows.

Brachial valve less convex than the pedicle and more compressed toward the extremities. The mesial fold is broad, low, rounded or angular, and sharply defined, with the plications more or less indistinct and irregular in arrangement. Aside from plications and lines of growth, the surface markings consist of rather coarse elongate papillae that are higher and thicker on their anterior portions than on their posterior. These papillae show but slight trace of linear arrangement, being scattered irregularly over the plications and in the furrows. No trace of striae is visible.

Remarks. — This form resembles in its general shape the young of the typical *whitneyi*, considerably broadened along the hinge-line. In general thinness of shell, regularity of convexity, lack of cardinal productions of the hinge-line, and absence of heavy growth lines, this variety differs from *S. whitneyi*. Moreover, the papillae differ so greatly from those of *whitneyi* that one would almost be justified in considering *subsiduus* as a distinct species.

Occurrence. — Throughout the middle portions of the Spirifer zone, and probably in the upper as well. Types are from Rockford and Hackberry Grove.

Holotype. — No. 8143, University of Michigan; *Paratypes.* — Nos. 26071 and 26072, Walker Museum.

SPIRIFER WHITNEYI PRODUCTUS C. L. Fenton

(Plate XXVIII, Figs. 17–22; Plate XXXI, Figs. 1–4)

Spirifer Whitneyi productus C. L. Fenton, Am. Mid. Nat., vol. 5, p. 220, pl. 6, figs. 5–6. 1918.
Spirifera whitneyi var. *disparilis* Webster, Contrib. to the Paleontology of the Hackberry Group, p. 3, pl. 2, figs. 5–8. 1906.

Description. — In measurements, character and arrangement of plications, and in surface-markings, this variety closely conforms to *S. whitneyi*. The plications may differ slightly from those of the typical form in a somewhat greater coarseness, but this difference is not distinctive. The distinctive feature of the variety *productus* is the development of sharp mucronations on the cardinal extremities. Even this character is more or less unstable, and it is not uncommon to find a specimen in which one extremity is produced and the other almost normal. In the specimens from the Strophonella beds, however, there is a fairly regular and uniform development of the mucronations, accompanied by a noticeable coarseness of the plications.

Remarks. — The specimen figured by Webster as *S. whitneyi disparalis* is a very good example of the variety *productus*. Webster's name had to be abandoned because it was preoccupied by Hall in 1857.

Occurrence. — Throughout the middle and upper portions of the Spirifer zone, but particularly in the Strophonella faunule.

Holotype. — No. 26022; *Plesiotypes.* — Nos. 26023, Walker Museum, and 8144, University of Michigan, and specimen in the collection of C. L. Webster.

SPIRIFER WHITNEYI OWENENSIS Webster & Fenton, n. var.

(Plate XXX, Figs. 20–22)

Description. — Shell of large size, with greatest width anterior to the hinge-line. Dimensions of holotype and paratype: length of pedicle valve, 22.6 mm. and 17.5 mm.; length of brachial valve, 23 mm. and 19.3 mm.; width, 32.4 mm. and 27.3 mm.; thickness, 21.2 mm. and 18.3 mm.; height of cardinal area, 10.4 mm. and 8.4 mm.

Pedicle valve subpyramidal, with high umbonal region, pointed, slightly incurved beak and abruptly sloping, flattened lateral slopes. The mesial sinus is broad and sharply defined, but very shallow and flat-bottomed and even reversed, somewhat as in *S. whitneyi rockfordensis*. The extension is broad and subrectangular rather than lingual. Plications broad, low, and rounded, numbering 20 to 25 on the lateral slopes and 7 to 10 on the sinus.

Brachial valve moderately convex, with umbonal region prominent and beak short and incurved. Antero-lateral slopes curve abruptly to the margin; postero-lateral ones compressed near the extremities. Mesial fold sharply defined, but broad, and low except in the anterior third of gerontic specimens. Plications correspond to those of the opposite valve. Valves marked by numerous strong growth lines, and by finer markings similar to those of. *S. whitneyi.*

Remarks. — This species represents a further development of certain forms of *whitneyi*, with high area and flat sinus, that are common in the upper portions of the Spirifer zone. Its diagnostic characters are the high, flattened, vertically striate area, the subpyramidal pedicle valve with sharp, horizontally directed beak and the flat, squarely produced sinus and fold.

Occurrence. — Shaly beds of the Idiostroma zone of the Owen, mainly at Owen's Grove; also from the Brachiopod faunule at Hackberry Grove.

Specimen. — Collection of C. L. Webster.

SPIRIFER cf. WHITNEYI OWENENSIS

Description. — Shell of large size, with greatest width anterior to the hinge-line. Dimensions of cast: length of pedicle valve, 18.5 mm.; length of brachial valve, 20.6 mm.; width, 34.4 mm.; thickness, 21.5 mm.; height of cardinal area, 11 mm.

This single cast, from the dolomitic beds of the Owen, exhibits the high area, broadly subpyramidal pedicle valve and projecting brachial beak of the form *owenensis*. It differs from it, however, in a broad, rounded sinus and shallow brachial valve, with flat antero-lateral slopes. The plications are similar to those of *owenensis*.

Specimen. — Collection of C. L. Webster.

SPIRIFER RARUS Webster

(Plate XXX, Figs. 23–26; Plate XXXI, Figs. 12–15)

Spirifera rara Webster, Contrib. to the Paleontology of the Hackberry Group, p. 4, pl. 2, figs. 12–15. 1906.

Description. — Shell of medium or large size, with greatest width along or anterior to the hinge-line. Dimensions of the plesiotypes: length of pedicle valve, 13.5 mm. and 24.3 mm.; length of brachial valve, 15.8 mm. and 24.4 mm.; width, 21.4 mm. and 29.6 mm.; thickness, 17.7 mm. and 27.4 mm.; height of cardinal area, 9.6 mm. and 13.4 mm.

Pedicle valve subpyramidal, with sharp and slightly incurved beak. Lateral slopes, which are flattened, descend abruptly to the margin. Area high and vertically striate; delthyrium narrow and high. Mesial sinus broad, shallow, and sharply defined, with a moderate lingual extension anteriorly. Plications rounded and separated by deep, narrow furrows; fine surface markings unknown. Brachial valve moderately convex, with umbo produced sharply beyond the cardinal margin. Mesial fold high, rounded and sharply defined.

Remarks. — This species bears a certain resemblance to *S. whitneyi owenensis* in the high area and subpyramidal form of the pedicle valve. It differs from that form, however, in the greater development of both characters. Moreover, the delthyrium is notably narrower and the sinus and fold more curved than in *owenensis*. From *S. whitneyi* this species differs in most of the points mentioned above, as well as in lesser proportional width, and the strong tendency of the brachial valve to be the longer, which is due to the sharp umbonal projection.

Occurrence. — Middle and upper Spirifer zone and lower Owen.

Types. — Collection of C. L. Webster.

SPIRIFER ROBUSTUS Webster

(Plate XXXI, Figs. 9–11)

Spirifera robusta Webster, Contrib. to the Paleontology of the Hackberry
Group, p. 9, pl. 2, figs. 9–11. 1906.

Description. — Shell large, wider than long, with the greatest
width along the hinge-line. Dimensions of the holotype: length
of pedicle valve, 23 mm.; length of brachial valve, 19 mm.;
width, 27 mm.; thickness, 21 mm.; height of cardinal area
about 2.5 mm.

Pedicle valve moderately convex, the greatest convexity
being near the middle; lateral and antero-lateral slopes rounded;
postero-lateral slopes somewhat compressed. Mesial sinus
broad, deep, rounded, and extending to the beak; it is produced anteriorly to form a rounded lingual extension. Beak
heavy, incurved; area arched; delthyrium broadly triangular.
Brachial valve less convex than the pedicle, with the anterior
slopes abrupt. Fold high, sharply defined, rounded, and produced to correspond to the extension of the sinus, but indefinite
near the beak. Surface on either side of the fold and sinus
marked by 4 or 5 strong, rounded plications and 2 or 3 weak
ones, leaving a considerable space near the cardinal extremities
without plications. Sinus marked by 4 or 5 weak, rounded
plications, those on the fold corresponding. Both valves marked
by heavy lines of growth. Finer surface markings undetermined.

Remarks. — This shell is characterized by its gibbous form,
coarse, rounded plications, smooth areas near the cardinal extremities, and general coarseness of appearance. Its relationships have not been determined, but probably it belongs with
the group of *Spirifer orestes*.

Occurrence. — Rare, in the Spirifer zone, Hackberry Grove.

Holotype. — Collection of C. L. Webster.

SPIRIFER SUBORESTES Webster

(Plate XXXI, Figs. 5–8)

Spirifera suborestes Webster, Contrib. to the Paleontology of the Hackberry Group, p. 8, pl. 2, figs. 16–18. 1906.

Remarks. — The illustration of this species by Webster is here reproduced. It is one of a group of several pustulose shells derived from *Spirifer orestes* H. & W.; its exact relations are not determined because the types are not available for study. Inasmuch as this entire group of *Spirifers* is the subject of a detailed genetic study made by the senior author, it seems undesirable to insert descriptions of any of the forms here. The description may be invalid, since it does not admit of satisfactory determination.

Genus PLATYRACHELLA Nov.

Genotype: **SPIRIFER MACBRIDEI** Calvin

Description. — Shell spiriferoid, small to large, with high and nearly flat cardinal area. Surface marked by strong plications, which may be either fine or coarse. Diagnostic characters are the presence of a well-defined delthyrial plate and impunctate shell structure. The former separates the genus from *Spirifer*, and the latter from *Pseudosyrinx*, with the genotype being most like the latter genus in general appearance.

It seems probable that many, if not most, of the *Spirifer*-like shells possessing delthyrial plates should be referred to this new genus. Certainly, the assumption of Hall and Clarke to the effect that pustulose surface indicates punctate structure is not to be relied upon, for *P. macbridei*, which is strongly pustulose, does not show the slightest trace of punctae. That this is not a matter of preservation is shown by the fact that associated species, such as *Cyrtina iowaensis*, show the punctae very plainly. Apparently this new genus occupies a position ancestral to Weller's *Pseudosyrinx*, which possesses the plate very strongly developed, and coarse punctae. It, in turn, appears to be ancestral to *Syringothyris*, which is punctate, and possesses both delthyrial plate and syrinx.

Lack of time has prevented a general examination of *Platyrachella*-like species. One form, however, *Spirifer asper* Hall, from the Cedar Valley, very clearly belongs to this new genus.

PLATYRACHELLA MACBRIDEI (Calvin)

(Plate XXIX, Figs. 23–30; Plate XXXII, Figs. 8–16)

Spirifera Macbridei Calvin, Am. Journ. Sci., 3d ser., vol. 25, p. 433. 1883.
Spirifer macbridei Calvin, Bull. Lab. Nat. Hist. State Univ. Ia., vol. 2,
    p. 166, pl. 12, fig. 3. 1892.
Spirifer macbridii Hall and Clarke, Pal. N. Y., vol. 8, pt 2, pp. 29, 31, 39,
    pl. 25, figs. 9–16 (*not* figs. 17–19). 1893.

Description. — Shell large, spiriferoid, with greatest width along the hinge-line. Dimensions of three typical specimens: length of pedicle valve, 18.5 mm., 18.4 mm. and 22 mm.; width, 37.5 mm., 36.7 mm. and 39.9 mm.; thickness, 22.3 mm., 25 mm. and 23.2 mm.; height of cardinal area, 15 mm., 15.8 mm. and 15.3 mm.

Pedicle valve broadly subpyramidal, with the surface sloping steeply from the umbonal region to the lateral and anterior margins. Mesial sinus originates at the beak and extends down the valve as a broad, deep, sharply defined trough that is produced anteriorly into a rounded or pointed lingual extension. In the sinus is a broad, low elevation which originates 12 to 15 mm. from the beak; in two or three specimens there are indications of rudimentary plications on slopes, although these may be nothing more than very coarse striae. Beak small, pointed, very slightly incurved; area broad and very high, and in most specimens more or less distorted. It is flattened or slightly concave; the concavity is just below the beak in most specimens, and slopes anteriorly from the margin at an angle of about 30 degrees. Delthyrium about twice as wide as high; central region well defined, as in *Syringothyris* and the typical *Pseudosyrinx* of the Mississippian; triangular lateral regions one and a half to two times as wide as the base of the delthyrium, and marked by vertical striae and horizontal growth lines. Each slope bears from 10 to 15 coarse, rounded or subangular plications, which are considerably crowded near the cardinal extremities. Internally the hinge-teeth are supported by slightly diverging dental lamellae which extend forward almost half the length of the valve, the divergence being equal to or less than that of the margins of the sinus. Diductor scars distinct but not elevated. Delthyrial plate well developed, flattened or concave, extending about one fourth of the way from the beak to the cardinal margin.

Brachial valve subtrapezoidal in outline, with maximum convexity in the umbonal region.[23] Surface curves abruptly to the cardinal margin, gently so toward the anterior, and is flattened toward the cardinal extremities. Mesial sinus large, sharply defined, non-plicate, rounded or subangular; beak small and incurved. Lateral slopes bear simple, coarse, rounded plications which correspond to those of the opposite valve, and are separated by deep, angular furrows. Internal characters of the valve well shown in the figure.

Surface of shell marked by minute papillae which are arranged upon fine irregular striae. On the slopes these striae run more or less irregularly, ascending the slopes of the plications. On the sinus they are bunched, wavy, and more or

[23] This is true in most specimens, but some show the maximum thickness at about the middle of the valve.

less obscured by the papillae. On the fold the bunching is less pronounced and the papillae are less distinct than on the sinus. There also are large numbers of very fine lines of growth, which are interspersed with heavier, lamellose ones near the margins. Shell impunctate.

Remarks. — Originally described as *Spirifer*, and usually identified as such, this species is taken as the type of the new genus *Platyrachella*. In general appearance it is strongly *Pseudosyrinx*-like, but the impunctate shell prevents reference to that genus.

Occurrence. — *P. macbridei* appears in the Gypidula faunule and extends throughout the Spirifer zone. It is most common in the middle beds at Rockford.

Plesiotypes. — Nos. 8217 to 8221, University of Michigan.

PLATYRACHELLA PULCHRA, n. sp.

(Plate XXIX, Figs. 20–22)

Description. — Shell of medium size, wider than long, with the greatest width along the hinge-line. Dimensions of the holotype: length of pedicle valve, about 18.5 mm.; length of brachial valve, 16.4 mm.; width, about 28.5 mm.; thickness, 15.8 mm.; height of cardinal area, 8.5 mm.

Pedicle valve normally spiriferoid, with the shell sloping abruptly to the anterior margin and somewhat compressed toward the cardinal extremities. Umbonal region high; beak prominent, sharp, incurved. Mesial sinus broad, rounded, with a trace of a plication on each slope, but with the bottom smoothly rounded. Lateral slopes bear 11 to 14 broad, rounded, simple plications, those bordering the sinus being lower than the others. Furrows broad, rounded or subangular. Cardinal area high, concave throughout, marked by vertical striae and horizontal lines of growth. Delthyrium broadly triangular, but slightly higher than wide.

Brachial valve much less convex than the pedicle; beak short and incurved. Lateral and anterior slopes gently convex; postero-lateral slopes compressed near the slightly produced cardinal extremities. Mesial fold low, broadly rounded, and non-plicate. Plications of the slopes correspond to those of the opposite valve.

Surface marked by papillose striae and radiating lines of papillae, the latter appearing on the slopes. In the sinus the papillae predominate, but are not distinctly bunched. Lamellose growth lines appear near the margins, but the finer lines are indistinct or lacking.

Remarks. — The assignment of this species to the genus *Platyrachella* is based upon the external appearance only. The delthyrial plate is visible, though it does not extend lower than the beak; the dental plates may also be distinguished. In general the shell resembles *P. macbridei*, but may be distinguished from it by the more spiriferoid form, rounder plications, concave area, prominent, curved beak, less triangular sinus, lower, rounder fold, and more distinctly produced cardinal extremities.

Occurrence. — This rare species is restricted to the upper portions of the Spirifer zone. The holotype was collected from the uppermost Strophonella beds at Bird Hill.

Holotype. — No. 8222, University of Michigan.

PLATYRACHELLA CYRTINAFORMIS (Hall & Whitfield)
(Plate XXIX, Figs. 1–7)

Spirifera cyrtinaformis Hall and Whitfield, 23d Ann. Rep. N. Y. State Cab. Nat. Hist., p. 238, pl. 11, figs. 21–24. 1872.
(??) *Spirifera cyrtinaeformis* Whiteaves, Contrib. to Can. Pal., 1, p. 222. 1891.

Cyrtia cyrtiniformis Hall and Clark, Pal. N. Y., vol. 8, pt. 2, p. 42, pl. 25, figs. 26–32. 1893.

Spirifer cyrtinaformis Calvin, Ia. Geol. Surv., vol. 7, p. 167. 1897.

Description. — Shell small, wider than long with the greatest width at the hinge-line or a little anterior to it; cardinal extremities angular. The dimensions of Hall and Whitfield's type and plesiotype: length of pedicle valve, 12.3 mm. and 11.2 mm.; length of brachial valve, 12.8 mm. and 12.9 mm.; width, 14.8 mm. and 16.9 mm.; thickness, 12.4 mm. and 12.5 mm.; height of cardinal area, 9.7 mm. and 9 mm.; length of hinge-line, 14.8 mm. and 15.8 mm.

Pedicle valve subpyramidal; the surface slopes from the umbo to the lateral margins in nearly straight lines and curves gently to the anterior margin. Mesial sinus originates at the beak, is shallow and about 6 mm. wide at the anterior margin; lingual extension directed upward at an angle of about 85 degrees to the plane of the valve. Beak pointed, very slightly incurved; cardinal area about half as high as wide, nearly flat and marked by fine vertical striae. Delthyrium about three times as high as wide; closed near the apex by a small delthyrial plate that for a short distance is slightly convex and then directed downward. Mesial sinus bears 10 distinct rounded or subangular plications at the anterior margin. These bifurcate and even trifurcate posteriorly. Each lateral slope bears 17.

Brachial valve moderately convex, the greatest convexity in the umbonal region; surface convex rather abruptly from the umbo to cardinal margins and gently to the lateral margins and still more gently to the anterior margins. Slight depressions near the cardinal extremities from auriculations. Mesial fold originates in the umbonal region and anteriorly becomes moderately high and wide; bears 9 to 10 plications of the same character as those in the sinus. Lateral slopes have plications to correspond to those of the opposite valve.

Surface of both valves marked by fine, regular, radiating plications, which are separated by broad, shallow troughs. In adult specimens the number of plications ranges from 40 to 55. The finer surface markings consist of fine, wavy, nodose striae, whose direction generally parallels that of the plications.

Remarks. — This species, described by Hall and Whitfield as a *Spirifer*, commonly has been referred to *Cyrtia*, despite its lack of the essential *Cyrtia* characters. It is referred to the genus *Platyrachella* on the strength of the delthyrial plate and impunctate shell.

Occurrence. — Throughout the Spirifer zone; most abundant in the Strophonella beds; rare in the Owen substage.

Plesiotypes. — Nos. 8207 to 8209, University of Michigan.

PLATYRACHELLA CYRTINAFORMIS HELENAE C. L. Fenton

(Plate XXIX, Figs. 8–10)

Spirifer cyrtinaformis helenae C. L. Fenton, Am. Mid. Nat., vol. 5, p. 216, pl. 6, figs. 11–17. 1918.

Description. — Shell small, wider than long with the greatest width along the hinge-line. Cardinal extremities angular. The dimensions of the holotype and paratype: length of pedicle valve, 11.8 mm. and 5.2 mm.; length of brachial valve, 12.3 mm. and 6.0 mm.; width along hinge-line, 20.0 mm. and 13.0 mm.; height of cardinal area, 9.8 mm. and 5.5 mm.

Pedicle valve from subtriangular to subtrapezoidal; the surface slopes in nearly straight lines from the umbo to the margins. Mesial sinus originates at the beak, becomes shallow and wide at the anterior margin where it is produced into a rounded lingual extension. The beak small and rounded; the high, flat cardinal area slopes anteriorly. Delthyrium about three times as high as wide; delthyrial plate short and directed inward. Surface of the valve marked by high, rounded, simple plications and deep rounded furrows; 29 to 36 plications, 5 to 7 of which lie in the sinus.

Brachial valve more convex than the pedicle; from the umbonal region the surface curves abruptly to the cardinal margin on each side of the beak and much less so to the anterior margins; postero-lateral slopes flat. Mesial fold not discernible from the general surface except near the anterior margin, where it is low and broad. The plications are similar in form and number to those of the opposite valve. Finer surface markings identical to those of *P. cyrtinaformis.*

Remarks. — This variety is characteristic of the uppermost portions of the Spirifer zone, and particularly of the Strophonella beds at Bird Hill. It is worthy of note that in the two closely related species, *P. cyrtinaformis* and *P. alta,* there are broad varieties, and that these broad varieties have considerable resemblance — more, indeed, than the parent forms. Moreover, the *Spirifers* of this same zone go through a similar modification, in one case with complications so great as to very nearly baffle attempts at identification.

Occurrence. — Upper part of the Spirifer zone.

Holotype. — No. 26073; *Paratype.* — No. 26074, Walker Museum.

PLATYRACHELLA ALTA, n. sp.

(Plate XXIX, Figs. 11–13)

Description. — Shell small, wider than long, with the greatest width along the hinge-line. Dimensions of the holotype: length of pedicle valve, 12.6 mm.; length of brachial valve, 13.8 mm.; width, 15.6 mm.; thickness, 18.3 mm.; height of cardinal area, 12 mm.

Pedicle valve subpyramidal; beak pointed, but only slightly incurved. Lateral slopes steep and flattened; mesial sinus broad, rounded and lingually produced. Area triangular, almost flat, with a slight concavity near the beak. Delthyrium more than twice as high as wide; delthyrial plate small, restricted to the region of the beak. Area marked by vertical striae and horizontal growth lines. Sinus bears from 4 to 8 plications, which increase by bifurcation and implantation. The number near the beak is 2 or 3; that at the margin of the holotype, 7. Lateral slopes bear 10 to 14 coarse, rounded, simple plications, which are much crowded toward the extremities.

Brachial valve much less deep than the pedicle. It is moderately convex in the umbonal region, with narrow, horizontal area, short, sharp beak, and steep, flattened slopes. The sinus is indistinct in the umbo, and high though imperfectly defined anteriorly. The plications correspond to those of the opposite valve, but are somewhat more angular.

The interiors of the valves are unknown. Presumably they closely resemble those of the related species *cyrtinaformis*. The surface-markings, so far as they are preserved, appear to resemble those of that species.

Remarks. — This species is closely allied to the preceding, differing from it mainly in the character and number of plications. Also, it appears somewhat higher up in the formation than does *P. cyrtinaformis*.

Occurrence. — Throughout the upper half of the Spirifer zone.

Holotype. — No. 8213, University of Michigan.

PLATYRACHELLA ALTA LATIMARGINATA, n. var.

(Plate XXIX, Figs. 14–19)

Description. — Shell of small to medium size, wider than long. Dimensions of the holotype and allotype: length of pedicle valve, 12.5 mm. and 9.7 mm.; length of brachial valve, 15 mm. and 10.6 mm.; width, 27.6 mm. and 17.9 mm.; thickness, 13.8 mm. and 9 mm.; height of cardinal area, 9.1 mm. and 6.4 mm.

Description of Fossils 167

Pedicle valve broadly subpyramidal, with the lateral slopes noticeably depressed toward the extremities. Beak sharp and but slightly incurved. Area broad, high and vertically striate; width from four to six times that of the base of the pedicle opening. Plications of the sinus number 4 to 6, the middle one generally being dichotomous. Lateral slopes bear 10 to 12 — rarely 14 — strong rounded or flattened plications which are separated by broad, deep, rounded furrows.

Pedicle valve with sharp beak, slightly convex central region, and flattened or concave lateral slopes. The plications of the low, indistinct fold are finer than those of the slopes and tend to dichotomize. Plications correspond to those of the opposite valve. Surface markings much like those of *P. cyrtinaformis*, but less regular and coarser.

Remarks. — This variety occupies much the same position in the *P. alta* line as does *helenae* in the line of *P. cyrtinaformis*. The irregularity of shape, and coarseness of plications serve as reliable means of separation.

Occurrence. — Throughout the upper portions of the Spirifer zone, particularly in the Strophonella and Stromatoporella faunule at Bird Hill and Hackberry Grove.

Holotype. — No. 8214; *Paratype.* — No. 8215, University of Michigan.

Genus RETICULARIA M'COY

RETICULARIA INCONSUETA, n. sp.

(Plate XXX, Figs. 6–9)

Reticularia, n. sp. C. L. Fenton, Am. Journ. Sci., 4th ser., vol. 48, p. 372. 1919.

Description. — Shell below medium size to large with greatest width anterior to the hinge-line. Dimensions of holotype, allotype and somewhat damaged large specimen: length of pedicle valve, 18.7 mm., 16 mm. and 26 mm.; length of brachial valve, 16.5 mm., 15 mm. and 22 mm.; width, 22.8 mm., 19 mm. and 31 mm.; thickness, 15.5 mm., 11.7 mm. and 19 mm.; height of cardinal area, 3 mm., 2.7 mm. and 4.3 mm.

Pedicle valve rather strongly convex, greatest convexity posterior to the middle; surface curves abruptly from the umbonal region to the cardinal margins and more gently to the lateral and anterior margins; shell slightly depressed near the cardinal extremities. Mesial sinus originates near the beak, moderately deep, wide and rounded in the bottom. Beak very prominent, pointed and incurved; cardinal area uniformly arched in small specimens and the base flattened in large ones; cardinal extremities well defined and rounded; delthyrium open, broadly triangular; dental lamellae extend beyond the cardinal area. Sinus non-plicate; lateral slopes bear from 4 to 6 broad low, rounded plications which are poorly defined in young stages.

Brachial valve somewhat less convex than the pedicle; surface curves rather gently to the cardinal margins and still more so to the lateral and anterior margins, somewhat compressed near the cardinal extremities. Mesial fold originates in the umbonal region and becomes rather high, rounded and wide anteriorly, about 12 mm. at the anterior margin; less prominent in small specimens. The beak is small, rounded and slightly incurved; cardinal area about 1.5 mm. in height in large individuals, nearly horizontal in position. Mesial fold non-plicate; on lateral slopes are plications corresponding to those of the opposite valve.

Surface of both valves marked by concentric growth lamellae, from 7 to 20 in the space of 5 mm. These are beset with fine spines, about 5 to the space of 1 mm.

Remarks. — This large *Reticularia* has most commonly been referred to the species *fimbriata*. The differences between the two species are so great, however, as to make a contrast of them unnecessary.

Occurrence. — Throughout the upper portions of the Spirifer zone, but particularly in the Strophonella faunule at Bird Hill.

Holotype. — No. 8203; *Allotype.* — No. 8204; *Paratypes.* — Nos. 8205, University of Michigan, and 25882, Walker Museum.

CYRTINA IOWAENSIS, n. sp.

(Plate XXX, Figs. 1–5)

Cyrtina hamiltonensis Hall, Pal. N. Y., vol. 4, p. 268 (part). 1867.

Description. — Shell small, subpyramidal in form, wider than long with the greatest width along the hinge-line; cardinal extremities angular. The dimensions of two specimens, the narrow and wide forms: length of pedicle valve, 8.4 mm. and 9 mm.; length of brachial valve, 6.3 mm. and 7.3 mm.; width along hinge-line, 9.5 mm. and 12.3 mm.; height of cardinal area, 5.8 mm. and 7.8 mm.

Pedicle valve subpyramidal with the apex erect or moderately curved; the surface slopes steeply from the umbo to the lateral and anterior margins; lateral slopes slightly convex. Mesial sinus moderately deep; originates at the beak and is rounded or subangular in the bottom. Cardinal area high, usually flat in young individuals, but arched near the beak in the mature forms. The delthyrium high and narrow, covered by a convex pseudo-deltidium that is pierced by an elongate, narrow foramen, extending from the apex to one third or one half the distance to the cardinal margin. Cardinal area is marked by fine vertical lines near the delthyrium, coarser horizontal lines and punctae; pseudo-deltidium bears both growth lamellae and punctae. Each lateral slope bears from 4 to 7 simple rounded or subangular plications that originate along the cardinal margin; the plications bounding the sinus are heavier and more rounded. The wide form has a greater width, in proportion to the rest of the shell, along the hinge-line. It usually has one more plication with a flattened area at the cardino-lateral margin.

Brachial valve slightly convex. Mesial fold low and rather wide at the anterior margin; a slight depression present throughout the entire length or only toward the beak. The lateral slopes bear from 4 to 6 simple rounded or subangular plications; the first one adjacent to the fold is not infrequently as large as the fold itself; the furrow that separates this plication from the fold is rather deep, wide and rounded. The other plications become small toward the postero-lateral margins. Cardinal area linear. Internally the cardinal process is bifid and heavy; from it a pair of crural plates extend anteriorly, diverging but slightly; hinge-sockets excavated from the sides of crural plates. A prominent horizontal ridge is present between the crural plates near their extremities; it also extends laterally from them, though it is not as distinct. Median septum is a line originating at the horizontal ridge and extending a little beyond the mid-length of the shell.

Surface of both valves marked by lines of growth, particularly on the anterior margin. Shell punctate.

Remarks. — *C. iowaensis* has generally been referred to *Cyrtina hamiltonensis recta*, but differs from that form in the following respects: the cardinal area is not flat, nor are the plications sharply angular as are those in *recta*. The internal characters of the brachial valve are very different from those of *hamiltonensis* as shown in Fig. 44 on Pl. 44 in *Pal. N. Y.*, vol. 4. The cardinal process of *C. iowaensis* is heavy and the crural plates have a slight curvature and are less diverging than those of *C. hamiltonensis*. The conspicuous ridge between the crural plates seems to be absent in *C. hamiltonensis*.

Occurrence. — Throughout the Spirifer zone, and sparingly in the lower parts of the Owen. It is especially abundant in the Strophonella faunule at Bird Hill.

Holotype. — No. 7896; *Allotype.* — No. 7897; *Paratypes.* — Nos. 7898 and 7899, University of Michigan.

Phylum MOLLUSCA

Class PELECYPODA

Genus BYSSOPTERIA Hall

BYSSOPTERIA OCCIDENTALIS, n. sp.

(Plate XXXIII, Figs. 8-9)

Description. — Shell small, subtriangular, alate posteriorly but with body undefined. Length slightly greater than height Dimensions of holotype: length, 20 mm.; height, 18 mm.; thickness, 13.8 mm. Umbos very prominent; shell highly convex umbonally, flattened postero-cardinally, and compressed along the lower anterior portion. Shell marked by strong radiating ribs, rounded and wider than the interspaces, which involved the entire thickness of the shell in all but the umbonal regions.

Remarks. — This single cast is totally unlike anything else which we have seen from the upper Devonian of Iowa. So far as can be determined, it possesses the essential characters of Hall's *Byssopteria,* although much smaller than the typical *B. radiata* of the eastern Chemung.

Occurrence. — Spirifer zone. Very rare.

Holotype. — No. 26043, Walker Museum.

Genus PARACYCLAS Hall

PARACYCLAS SABINI White

(Plate XXXIII, Figs. 4–5)

Paracyclas sabini White, Proc. Acad. Nat. Sci. Phila., vol. for 1876, p. 31. 1876.

Description. — Shell small, subovate in outline. Dimensions of two typical specimens: length, 19.4 mm. and 13.4 mm.; height, 13.8 mm. and 11.2 mm.; thickness, 9.8 mm. and 6 mm. Other specimens, apparently distorted in growth or by pressure, show the proportion of height to length to be greater than that indicated here.

"Shell sublenticular, . . . beaks small, approximate, pointing forward, elevated little if any above that portion of the dorsal margin which lies behind them, but considerably above that portion in front of them; dorsal, posterior, and basal margins forming nearly one uniform curve, but the prominent front, which is the narrowest and thinnest part of the shell, has its margin more abruptly rounded; ligament small, slightly prominent, but is made apparently more prominent by two distinct, moderately deep, narrow grooves, one on each side of it, which extend from the beaks backward, and become obsolete upon the postero-dorsal region; valves broadly and nearly uniformly convex, the surface being marked by ordinary lines and slight undulations of growth." — White, 1876.

Remarks. — Dr. White's description was drawn from specimens of the type of the more elongate one figured, which are the most numerous. Other specimens, in which the proportions are more like those of *P. elliptica* Hall, are common.

Occurrence. — Common throughout the Spirifer zone.

Plesiotypes. — Nos. 8057 and 8058, University of Michigan.

PARACYCLAS VALIDALINEA Webster

(Plate XXXIII, Figs. 1-2)

Paracyclas validalinea Webster, Am. Nat., vol. 22, p. 1016. 1888.

Description. — Shell large, oblong, subelliptical in outline, with width greater than the height. Dimensions of the more complete plesiotype: length, 32.4 mm.; height, 28.7 mm.; thickness, 12.8 mm. Valves irregularly lenticular, with convexity but slightly increased in the umbonal region; umbos prominent where preserved, directed anteriorly; lower than the margin behind them. Hinge-line nearly straight; posterior margin nearly at right angles with it. Muscle scars very strong, their shape being well shown in the illustrations. Shell marked by several broad, radiating folds, and two deep impressions anterior to each umbo. Pallial line well marked.

Remarks. — This species, which may with some doubt be retained in the genus *Paracyclas*, is distinguished by its very strong and uniquely shaped muscle scars. It is known only from casts, which generally are badly weathered.

Occurrence. — Common throughout the Spirifer zone; uncertain in the Owen.

Plesiotypes. — No. 26028, Walker Museum.

PARACYCLAS PARVULA, n. sp.

(Plate XXXIII, Fig. 3)

Paracyclas elliptica C. L. Fenton, Am. Journ. Sci., 4th ser., vol. 48, p. 373. 1919.

Description. — Shell small, subcircular in outline, with height equal to or greater than the length; pallial margin excessively curved in the mid-region of the shell. Valves lenticular, convex medially and compressed toward anterior and posterior margins.

Umbos high and sharp; hinge-line straight. Pallial line marked by a row of depressions. Shell unknown. Length of holotype, 9 mm.; height, 9.3 mm.

Remarks. — The shape and small size of this species are characteristic. It generally has been referred to *P. elliptica* Hall, without any special reason for the identification.

Occurrence. — Throughout the Spirifer zone, but particularly in the upper portions. Probably in the Owen.

Holotype. — No. 8049, University of Michigan; *Paratype.* — No. 25816, Walker Museum.

PARACYCLAS SPP.

(Plate XXXIV, Figs. 1–2)

Among the fossils of the Owen substage are a number of large pelecypods, apparently of the genus *Paracyclas*, but too poorly preserved for identification. The accompanying illustrations show the shape and markings of two of these casts.

Figured specimens. — No. 21021, Walker Museum.

Genus GRAMMYSIA DeVerneuil

GRAMMYSIA (?) DUBIA, n. sp.

(Plate XXXIII, Fig. 10)

Description. — Shell large, subrectangular in outline. Dimensions of holotype: length, 29 mm.; height, 25.7 mm.; thickness, 17 mm. Valves moderately convex, becoming more so in the umbonal region, and flattened posteriorly. Umbos prominent, opposed, directed forward. Ligament prominent; hinge-line prominent, straight; elevated to the height of the umbos posteriorly. Pallial line prominent. Anterior adductor scars large, circular, but not prominent.

Remarks. — This species is characterized, among Hackberry pelecypods, by the subrectangular shape, great proportion of thickness to length, and weak muscle scars. The reference to *Grammysia* is only provisional.

Occurrence. — Spirifer zone.

Holotype. — No. 8068, University of Michigan.

Genus PTERINEA Goldfuss

PTERINEA HUSSEYI, n. sp.

(Plate XXXIII, Fig. 6)

Aviculopecten sp. undet. C. L. Fenton, Am. Journ. Sci., 4th ser., vol. 48, p. 373   1919.

Description. — Shell large, subrhomboidal, oblique; body subovate, narrow above; length and height about equal. Margins rounded, and produced postero-basally. Dimensions of the holotype: length, 30 mm.; height, 28.5 mm. Left valve moderately convex; right unknown. Wing apparently broken off. Hinge-line straight; produced anteriorly to form the ear. Umbo prominent, directed anteriorly. Surface marked by strong, radiating rays distant 2.5 to 4 mm. from crest to crest; between each pair of these are 2 to 4 finer rays, the distinction between the two grades decreasing toward the wing.

Remarks. — The reference of this species to the genus *Pterinea* rather than to any of the *Pectinidae* is based on shape and surface-markings alone, the hinge being unknown. The shell is not well preserved.

Occurrence. — Middle portions of the Spirifer zone.

Holotype. — No. 26038, Walker Museum.

PTERINEA (?) SP.

(Plate XXXIII, Fig. 7)

Among the collections are several shell fragments, including both interiors and exteriors, of left valves of a large shell probably belonging to *Pterinea*. The length of the largest specimen is about 50 mm.; the width about 47 mm. They appear to indicate a species somewhat like *Pterinea chemungensis* Hall. The surface-markings consist of radiating ribs separated by flattened spaces 1 to 2 mm. in width, and seemingly non-striate. Wings are not preserved.

Occurrence. — Middle portion of the Spirifer zone, mainly at Rockford.

Figured specimen. — No. 26038, Walker Museum.

Class GASTROPODA

Genus BELLEROPHON Montfort

BELLEROPHON SP.

(Plate XXXIII, Figs. 12–13)

Description. — Shell small to medium, closely enrolled, with body whorl large. Aperture large, somewhat expanded; umbilicus small and deep. A few casts show the slit in the outer lip, but none preserve the slit band. In most of the specimens the shell is lacking; where found it is so badly worn as to be quite smooth.

Occurrence. — Casts of this, or related species, too poor to permit identification beyond that of the genus, are common throughout the Spirifer zone, and less so in the Owen.

Figured specimen. — No. 26031, Walker Museum.

Genus PLEUROTOMARIA Sowerby

PLEUROTOMARIA (?) VERTICILLATA Webster

(Plate XXXVII, Figs. 11–12)

Pleurotomaria (?) verticillata Webster, Iowa Naturalist, vol. 2, p. 2. 1906.

Description. — "Shell subturbinate; spire elevated; height and width unequal; volutions three, expanding quite regularly to within about three-fourths of a volution of the aperture where it broadens out; body volution flattened or very broadly rounded above and with a corresponding curvature of the surface below; umbilicus of medium size; suture very distinct, and distant on the body whorl some distance back from the aperture. Surface unknown." — Webster, 1906.

Remarks. — We have not identified this species, nor have we seen Webster's types. This description is reprinted because of the opportunity to publish, seemingly for the first time, Webster's figures, and carries no opinion as to the validity of the description.

Occurrence. — Spirifer zone, Hackberry Grove.

Cotypes. — Collection of C. L. Webster.

Genus STRAPAROLLUS Montfort

STRAPAROLLUS CYCLOSTOMUS PORTLANDENSIS, n. var.

(Plate XL, Figs. 14–15)

Description. — Shell discoid in outline, of medium to large size, with spire moderately elevated. Whorls generally 4, the body whorl being much larger than the others. Umbilicus very large, but not deep because of the lowness of the spire. Aperture circular, its diameter equalling about two thirds that of the umbilicus. Surface marked by heavy wrinkles and finer striae of growth.

Remarks. — This variety differs from the typical *S. cyclostamus* of the Cedar Valley in its larger size, coarser striae, and looser expression. The sutures are more open, the body whorl larger in proportion to the rest of the shell, and the spire higher and sharper than in the Cedar Valley species. It seems probable that specimens with the shell fairly complete will show the Hackberry form to be a distinct species, but at present it seems desirable to consider it merely a variety of *cyclostamus* of Hall.

Occurrence. — Throughout the Spirifer zone, particularly in the upper beds at and near Hackberry Grove.

Holotype. — No. 26037; *Paratype.* — No. 20981, Walker Museum.

STRAPAROLLUS ARGUTUS, n. sp.

(Plate XXXVIII, Figs. 12–14)

Description. — Shell discoid in outline, spire very low. Whorls 3 to 4, the body whorl being much larger than the others, and broadening notably toward the aperture. Aperture always greater in diameter than the umbilicus. Surface unknown.

Remarks. — The distinguishing features of this species are the low spire, very large body whorl, and the small, shallow umbilicus. The last feature alone will serve to distinguish the species from the form *portlandensis*, which invariably has the umbilicus greater in diameter than the aperture. The same is true of the typical *cyclostamus* of the Cedar Valley.

Occurrence. — Spirifer zone; uncommon in the Owen.

Holotype. — No. 7890; *Paratypes.* — Nos. 8044, 8045, University of Michigan, 26039 and 26040, Walker Museum.

STRAPAROLLUS CIRCINATUS, n. sp.

(Plate XXXVIII, Figs. 4–5; Plate XLV, Figs. 4–5)

Platyschisma (?) McCoyi Webster, Contrib. to the Paleontology of the Hackberry Group, p. 7, pl. 2, fig. 20. 1906.

Description. — Shell small, circular in outline, more or less trochiform. Spire moderately elevated; whorls 4 or 5, expanding gradually from the aperture. They are flattened above, and bear a strong subangular ridge or carina. Outer lip strong. Under side of body whorl moderately convex, with some indications of a second, lower, carina. Surface marked by fine growth lines which curve backward from the suture, then forward, and finally, on the under surface, backward again. Umbilicus moderately broad and deep.

Remarks. — While we do not have Webster's specimen, there seems to be but little doubt that his *Platyschisma mccoyi* is a typical member of this species, probably made more angular by pressure. Specimens with the shell are rare, but there is an abundance of casts, such as the one figured, that probably belong to this species. Many of them are flattened, but others preserve the original elevation of the spire.

Occurrence. — Throughout the Spirifer zone; sparingly in the lower Owen.

Holotype. — No. 8046; *Paratypes.* — Nos. 8047 and 8048, University of Michigan.

Genus CYCLONEMA Hall

CYCLONEMA (?) PERSTRIALIS, n. sp.

(Plate XLI, Fig. 4)

Description. — Shell conical, with high spire. Whorls 3 to 4, enlarging gradually to the body whorl, which is moderately ventricose. Height of shell greater than the diameter of the body whorl. Whorls flattened on the periphery. Umbilicus quite covered by the reflected inner lip. Surface bears strong carinae, of which there are 2 on the upper surface and 3 on the lower surface of the body whorl. Transverse striae of growth probably are present, but fail to show in the specimen.

Remarks. — The actual generic relationships of the one specimen on which this species is based are doubtful. It may belong to *Cyclonemina* rather than *Cyclonema*. Among the gastropods of the Hackberry it is characterized by the very heavy carinate shell.

Occurrence. — Spirifer zone; rare.

Holotype. — No. 26042, Walker Museum.

Genus TURBO Linnaeus

TURBO (?) SPICULATUS, n. sp

(Plate XXXIII, Figs. 14–15)

Description. — Shell large, subovate, with slightly elevated spire. Whorls about 3, expanding rapidly, the body whorl making up more than half the volume of the entire shell. Aperture subcircular; inner lip heavy. Shell marked by slight folds, somewhat like those of *Floydia concentrica,* but less strong. The surface bears several hollow, spine-like projections that extend backward from the aperture. Surface markings obscured by a stromatoporoid.

Remarks. — While there must be some uncertainty in the founding of a species of gastropod on a single specimen whose surface-markings are so hidden as to be undeterminable, we feel that to refer this form to any species already described would be an error, and that to mention it without name would be unnecessary caution. The shape of the low-spired shell, the large body whorl, and the backwardly directed, hollow projections all appear to be distinctive.

Occurrence. — Spirifer zone; found only at Hackberry Grove, where it is rare.

Holotype. — No. 26041, Walker Museum.

GENUS PYCNOMPHALUS LINDSTRÖM

PYCNOMPHALUS KINDLEI, n. sp.

(Plate XXXVIII, Fig. 16)

Description. — Shell medium to large, subcircular in outline, with rounded, convex upper surface. Whorls 3 to 5 — generally 4 in adult specimens. They expand gradually from the apex, the body whorl occupying about one half the total diameter of the coil. Suture close; whorls regular and rounded. Aperture circular. Umbilicus, which shows well in the casts, is deep. Surface unknown.

Remarks. — This species, whose cast appears to correspond in all essentials to *Pycnomphalus* Lindström, known mainly from Europe, is one of the more common gastropods of the Hackberry. The rounded, regular whorls, and the deep umbilicus, are characteristic of all specimens from which the matrix can be cleaned.

Occurrence. — Throughout the Spirifer zone, being specially common at Rockford; throughout the Owen.

Holotype. — No. 7883, University of Michigan; *Paratypes.* — Nos. 25813, 25748 and 25837, Walker Museum.

GENUS FLOYDIA WEBSTER

Genotype: **FLOYDIA CONCENTRICA** Webster

Floyda Webster, Iowa Naturalist, vol. 1, p. 39. 1905.
Floydia C. L. Fenton, Am. Mid. Nat., vol. 5, p. 221, pl. 7, figs. 4–5. 1918.

Description. — Shell large and thick; spire high, or low with the whorls partly enrolled upon each other. Whorls rounded or more commonly flattened on the side; they number 4 to 6 — rarely 7 — and show the outer one to be but moderately expanded. Suture well marked and deep, but closed; umbilicus lacking. Aperture large, subcircular to subovate, and generally distorted. Outer lip thin; inner lip thick and more or less flattened, as is also the ventral portion of the body whorl. Shell smooth or marked by heavy wrinkles and folds, as well as finer striae.

Remarks. — The members of this genus commonly have been assigned to *Naticopsis* McCoy. From this genus *Floydia* differs in a characteristically higher spire and smaller body whorl. The columella is exceptionally heavy, and the typical wrinkles and folds appear to be lacking in *Naticopsis*.

Occurrence. — At present the genus is known only from the Hackberry stage, where it ranges from the base of the Spirifer zone to the top of the Owen. It seems probable that several species from Cedar Valley rocks may be referred to *Floydia* upon reëxamination.

FLOYDIA GIGANTEA (Hall & Whitfield)

(Plate XXXIV, Fig. 5; Plate XXXV, Figs. 1–10; Plate XXXVI, Fig. 1;
Plate XXXVII, Fig. 10)

Naticopsis gigantea Hall and Whitfield, 23d Ann. Rep. N. Y. State Cab.
Nat. Hist., p. 238, pl. 12, figs. 8–10. 1873.
Naticopsis magnifica Webster, Iowa Naturalist, vol. 1, p. 58. 1905.

Naticopsis gigantea Webster, Iowa Naturalist, vol. 2, p. 45, pl. 9, figs. 1–11. 1909.
Naticopsis gigantea C. L. Fenton, Am. Mid. Nat., vol. 5, p. 222, pl. 7, figs. 1–3. 1918.

Description. — Shell medium to very large in size, thick, ovate-conical, with spire moderately to sharply ascending. Whorls somewhat angular, flattened on the sides and below, with strong, deep sutures. Body whorl occupies one half to two thirds the entire height of the shell. Aperture large, oblique, slightly extended below. Columella heavy and flattened on the lower half; in the upper the lip becomes rounded, but remains quite distinct from the whorl. Shell marked by fine lines of growth and coarser wrinkles of the same origin.

Remarks. — This is a highly variable species, particularly in the height of spire. Some of the specimens are much flattened by pressure, and a few appear to have been elevated by the same means. Even in undeformed specimens, however, there is great variation, many showing a low spire and large, round body volution. Of these, the forms with moderately high spires are here considered as being typical of *F. gigantea*, while those of more than ordinary height, although not typical, are yet referred to the species. For the low forms, with large body whorl, Webster's name *hackberryensis* is used.

From its associate, *F. concentrica*, this species differs in the generally greater height of spires and the lack of well-defined shell folds. There are, it is true, strong wrinkles on many of the specimens, but they do not develop into folds affecting the whole shell, as in the genotype.

Occurrence. — Throughout the Spirifer zone and the Owen substage.

Plesiotypes. — No. 25807, Walker Museum, and specimens in the collection of C. L. Webster.

FLOYDIA GIGANTEA HACKBERRYENSIS (Webster)

(Plate XXXII, Figs. 3–4; Plate XL, Fig. 16)

Naticopsis hackberryensis Webster, Iowa Naturalist, vol. 2, p. 3. 1906.
Naticopsis hackberryensis Webster, Iowa Naturalist, vol. 2, p. 46, pl. 9, figs. 14–15. 1909.
Naticopsis gigantea hackberryensis C. L. Fenton, Am. Journ. Sci., 4th ser., vol. 48, p. 373. 1919.

Description. — Shell of medium size, with 3 to 4 volutions, which are rounded or oval in section. The suture is deep, the spire low, and the body whorl very large. The aperture, which is poorly represented in the casts, appears to have been oblique and nearly circular; there is a slight indication of a false umbilicus. The outer lip appears to have been slightly reflexed.

Remarks. — This variety, which is known only from casts, possesses the general shape of casts of *F. concentrica*, but lacks the heavy folds. From *F. gigantea* it differs in the low spire and large, slightly flattened body whorl.

Occurrence. — Throughout the Spirifer zone; common in the Idiostroma zone, and less so in the upper Owen.

Cotypes. — Collection of C. L. Webster; *Plesiotype,* No. 8036, University of Michigan.

FLOYDIA CONCENTRICA Webster

(Plate XXXIV, Fig. 6; Plate XXXVII, Figs. 4–9; Plate XLIII, Figs. 1–3)

Floyda concentrica Webster, Iowa Naturalist, vol. 1, p. 39. 1905.
Floydia concentrica C. L. Fenton, Am. Mid. Nat., vol. 5, p. 221, pl. 7, figs. 5–5A. 1918.

Description. — Shell large, depressed-conical, spire moderately high, high, or even low. Whorls 4 to 6, flattened outwardly; body whorl larger than the rest and more flattened. Sutures deep, but closed. Columella heavy and flattened; outer lip very weak, being preserved only where protected by growths of stromatoporoid. Body whorl, and 1 or 2 of the others, marked by heavy, deep folds, which arch obliquely backward and then forward from the suture. These folds are specially strong on the body whorl, where they involve the entire thickness of the shell, and commonly show in the casts.

Remarks. — The folds, just described, are the most distinctive feature of this species. Other characters serving to distinguish it from *gigantea* are the lower and heavier spire (general), more wrinkled shell, and much lesser tendency to become crystalline in fossilization. This character appears to be quite distinctive, and the crumbly, crystalline nature of *gigantea* shells probably accounts for the fact that they seldom are found with more than fragments of the shell remaining — and these crumble if the specimens are allowed to soak.

Floydia concentrica, too, is notable because of the large number of endoparasitic forms which it harbors. *Aulopora*, generally *A. iowaensis*, *Tabulophyllum*, *Pachyphyllum*, *Hederella*, *Hernodia*, and other bryozoa, *Crania*, *Serpula*, and *Spirorbis*, are all among the genera commonly to be found on the shells of this species. Other species fail to show any such association, or any specific association, as appears to exist between *concentrica* and *A. iowaensis*.

Occurrence. — Hackberry stage, above the Gypidula faunule.

Holotype and several *Plesiotypes.* — Collection of C. L. Webster; another *Plesiotype.* — No. 25741, Walker Museum.

FLOYDIA CONCENTRICA MULTISINUATA C. L. Fenton

(Plate XXXIV, Fig. 7; Plate XXXVIII, Fig. 15)

Floydia concentrica multisinuata C. L. Fenton, Am. Mid. Nat., vol. 5, p. 223, pl. 7, fig. 4. 1918.

Description. — Shell large and thick, with general shape like that of *F. concentrica*. Body whorl tends to be somewhat larger and more flattened than in the primary form, and the spire normally is high. The distinguishing feature of the variety is the frequency and strength of the shell folds. A typical specimen of *F. concentrica* shows 4 folds in the space of 35 mm.; the plesiotype of *F. concentrica multisinuata* shows 6. As is the case in the primary form, the folds tend to weaken above the body whorl, although in the holotype they are very strong throughout.

Occurrence. — Spirifer zone; uncertain in the Owen.

Holotype. — No. 25740, Walker Museum; *Plesiotype.* — No. 8037, University of Michigan.

GENUS WESTERNIA WEBSTER

Description. — Shell small to large, conical, with high spire; like *Loxonema* in general expression. Whorls numerous, generally 5 to 10; slightly flattened or gently rounded, and somewhat inflated below the center. Aperture subcircular to subovate, oblique; columella strong. Shell thick, marked only by fine lines of growth.

Remarks. — Webster's genus *Westernia* was founded on the group of forms which he had described as *Eoxonema gigantea*, *L. owenense* and *L. crassum*. In 1897 Calvin [24] expressed the opinion that these forms "belong with the peculiar form called

[24] *Ia. Geol. Surv.*, vol. 7, p. 166. 1897.

Naticopsis gigantea by Hall and Whitfield; and it is quite possible that this group will have to be assembled under a new generic description." Webster, however, did not believe that such union was warranted, and proposed two new genera, *Floydia* and *Westernia*.

In working over these species we attempted to find the relationship between *Loxonema* and *Westernia*, which in casts are very similar, and were forced to conclude that the shell of *Westernia* is sufficiently different from *Loxonema* to show the genera to be quite distinct. On the other hand, the shells of *Floydia* and *Westernia* are practically identical, and the difference in shape alone seems hardly enough to cause generic distinction. Moreover, casts are particularly difficult of identification, so that for practical as well as structural reasons we are inclined to consider Webster's group as a subgenus of *Floydia*, but lack sufficient information definitely to place it in that rank.

WESTERNIA GIGANTEA Webster

(Plate XLIV, Figs. 4–5; Plate XLV, Figs. 1–2)

Loxonema gigantea Webster, Am. Nat., vol. 22, p. 445. 1888.
Westernia gigantea Webster, Contrib. Paleontology of the Hackberry Group,
    p. 2, pl. 2, figs. 1, 2; pl. 3, figs. 7, 8. 1906.

Description. — Shell large, elongate-conical, with blunt apex. Length of adult specimens from 10 cm. to 13.5 cm. Whorls 6 to 7, the body whorl being considerably enlarged. Suture strongly marked; aperture subcircular. Shell thick, like that of *Floydia*, with fine growth lines showing in unweathered specimens. The distinctive characters appear to be the large size and very high spire.

Occurrence. — Uncommon in the upper beds of the Spirifer zone, where it occurs only as casts; common throughout the

Owen at Hackberry Grove, Owen Grove, Rockwell, and other localities.

Types. — Collection of Clement L. Webster.

WESTERNIA GIGANTEA OWENENSIS Webster

(Plate XXXI, Fig. 16; Plate XLIV, Fig. 1–3; Plate XLV, Fig. 3)

Loxonema owenensis Webster, Am. Nat., vol. 22, p. 446. 1888.
Westernia owenensis Webster, Contrib. Paleontology of the Hackberry
    Group, p. 2, pl. 2, figs. 3, 4; pl. 3, fig. 2. 1906.

Description. — Shell large, adults measuring 14 to 16.5 cm. in height. Shape elongate-conical; whorls 8 to 10, oblique, rounded, and enlarged below the center. Suture deeply channeled; shell 2 to 6 mm. in thickness, apparently smooth.

Remarks. — Although it is described as a distinct species, there seems to be nothing about this shell which clearly separates it from *W. gigantea.* According to Webster the distinguishing characters are the great size, and the depth of the suture, but these do not seem to be more than varietal characters at best.[25]

Occurrence. — Owen substage.

Types. — Collection of C. L. Webster.

WESTERNIA PULCHRA, n. sp.

(Plate XXXIV, Fig. 4; Plate XXXVIII, Fig. 7)

Description. — Shell of medium size, elongate-conical. Average height about 30 mm., but one of the paratypes shows a height of 50 mm. Whorls 6 or 7, rounded but moderately;

[25] There may be some question as to the propriety of even this distinction; the description appears to be very imperfect. *Loxonema crassum,* described on the same page as *W. g. owenensis,* is quite unidentifiable from the description, and so must be dropped.

body whorl considerably larger than the rest. Aperture oblique, elongate-clliptical in outline, and constricted above. Sutures sharp and deep, but narrow. Columella lip somewhat extended. Apex sharp.

Remarks. — This shell generally has been referred to *Loxonema hamiltoniae* Hall, for no particular reason whatever. The shape of the whorls is very different from that of the eastern form, and no specimens have been found with the shell present. Since, for all that is known, there are no identifiable *Loxonemas* in the Hackberry, the presumption is that this shell belongs to *Westernia*, which it closely resembles in form. It differs from *W. gigantea* in the much smaller size, greater regularity of surface, more elliptical aperture, and fewer whorls.

Occurrence. — Throughout the Spirifer zone, particularly at Hackberry Grove. Rare in the Idiostroma zone; probably lacking above.

Holotype. — No. 7881; *Paratypes.* — Nos. 7882, University of Michigan, and 26030, Walker Museum.

GENUS TRACHYODOMIA MEEK & WORTHEN

TRACHYODOMIA PRÆCURSOR ROCKFORDENSIS (Webster)

(Plate XLV, Figs. 6–7)

Trochus (Paleotrochus) præcursor var. *rockfordensis* Webster, Contrib. to the Paleontology of the Hackberry Group, p. 7. 1906.

Trochus (Paleotrochus) præcursor Webster, ibid., p. 5, pl. 2, figs. 22, 23. 1906.

Description. — Shell small. Whorls 4 or 5, the body whorl making up about two thirds of the total height of the shell, which reaches 18 mm. Body whorl depressed below; aperture somewhat oblique. No definite carina. Outer lip simple; inner without callosity. Surface marked by strong tubercles arranged in spiral bands.

Remarks. — This form differs from the typical *T. præcursor* of New York in that it is larger, heavier in shell, without prominent carina, and is considerably flattened outwardly. The tubercles, too, are coarser than are those of the New York form. There appears to be no important difference between the large specimen, which Webster described as *præcursor*, and the small ones which he made the types of the variety *rockfordensis;* in fact, the larger specimen offers more contrast with the New York form than do the small ones.

Occurrence. — Spirifer zone, Hackberry Grove and Rockford.

Cotypes. — Collection of C. L. Webster.

Genus HOLOPEA Hall

HOLOPEA (?) IOWAENSIS Webster

(Plate XXXII, Figs. 1–2; Plate XXXVIII, Fig. 6)

Halopea (?) iowaensis Webster, Iowa Naturalist, vol. 2, p. 3. 1906.
Halopea (?) iowaensis Webster, Iowa Naturalist, vol. 2, p. 46, pl. 9, figs. 12, 13. 1909.

Description. — Shell small, conoid-globose; dimensions of plesiotype: height, 21.7 mm.; maximum diameter of body whorl, 13.2 mm. Whorls 3 to 4, rounded, and increasing rapidly in size. Suture strong and deep; aperture subcircular to subovate. Known only from casts.

Remarks. — Although Webster's types were not at hand for comparison, the photographed specimen was identified by him as belonging to the typical *iowaensis*. The generic reference is, of course, based on shape alone.

Occurrence. — Rather uncommon in the Spirifer beds.

Plesiotype. — No. 26034, Walker Museum.

GENUS DIAPHOROSTOMA FISCHER

DIAPHOROSTOMA INSOLITUM (Webster)

(Plate XXXIV, Fig. 3; Plate XXXVI, Figs. 2–4)

Platyostoma insolita Webster, Iowa Naturalist, vol. 2, p. 4. 1906.
Naticopsis gigantea var. *depressa* Webster, Iowa Naturalist, vol. 2, p. 3. 1906.

Description. — Shell large, subovate; spire moderately elevated above the body whorl; whorls 3 or 4, 3 being the number generally preserved. Body whorl much expanded, somewhat ventricose along the suture, and flattened or rounded outwardly. For 5 to 15 mm. back from the aperture it is rather broadly separated from the whorl above. Aperture nearly vertical, subovate; columellar lip apparently thickened and extended. Surface unknown.

Remarks. — This species is known only from casts. It is characterized by the very large aperture, rapidly enlarging body whorl, and generally low, flattened spire. To it we refer the casts which Webster called *N. gigantea* var. *depressa*, since they appear to be but little different from the typical *insolitum.*

Occurrence. — Cerro Gordo substage, particularly the middle portion of the Spirifer zone.

Cotypes. — Collection of C. L. Webster; *Plesiotype.* — No. 25843, Walker Museum.

DIAPHOROSTOMA ANTIQUUM (Webster)

(Plate XLIV, Figs. 6–7)

Platyostoma antiquis Webster, Iowa Naturalist, vol. 1, p. 59. 1905.

Description. — "Shell of medium size; subovate to semicircular in general outline when viewed from above; spire elevated above the body volution, and proportionately somewhat more so in the young than in the adult forms; volutions three; the outer volution expanding toward the aperture, and large and broad at the aperture; aperture large, circular to semi-circular or obliquely subovate, judging from the cast; the inner lip appears to have been reflexed in both young and old specimens. Surface markings unknown, as the species is known only from the cast." — Webster, 1905.

Remarks. — The writers have not seen specimens which they were able to identify with this species. This perhaps is due to the great similarity existing between a large number of gastropod casts in the Hackberry.

Occurrence. — Spirifer zone, at various localities.

Types. — Collection of C. L. Webster.

DIAPHOROSTOMA MODESTUM (Webster)

(Plate XXXVII, Figs. 1–3)

Platyostoma (?) modesta Webster, Iowa Naturalist, vol. 2, p. 2.   1906.

Description. — "Shell small; a somewhat oblong-ovate outline is presented when viewed from above; spire elevated above the body whorl. Outer volution expanding from the apex and quite strongly produced laterally at the aperture; aperture apparently subcircular to obliquely subovate. Upper surface of upper volutions somewhat flattened, with a narrow, somewhat flattened area along the suture of the body whorl.

"Surface — much of the shell is still preserved in the specimens — apparently smooth." — Webster, 1906.

Remarks. — Like the preceding species, this has not been identified in our collections. Probably some casts so closely resemble those of *Floydia gigantea* as to make separate identification of them impossible.

Cotypes. — Collection of C. L. Webster.

DIAPHOROSTOMA IRRASUM, n. sp.

(Plate XXXVIII, Figs. 8–9)

Description. — Shell small, subovate, with spire but slightly elevated above the body whorl. Whorls generally 3 in number, the outer one enlarging very rapidly. Upper surface of whorls somewhat flattened; outer portion strongly rounded. Aperture large, vertical, and subovate. Inner lip heavy and rounded; outer lip thin. Surface marked by coarse, wavy lines of growth, which grow stronger near the aperture.

Remarks. — This species has some appearance of being the young of *D. insolitum* (Webster). Careful comparison, however, shows it to have a higher rate of expansion than does that species, particularly in youth, a proportionately larger aperture, and too many whorls to correspond with young stages of *D. insolitum*. From *D.* (?) *mirum* and *D.* (?) *pervetum* Webster, of the Striatula zone it differs in having a strongly striate shell.

Occurrence. — Upper portions of the Spirifer zone, particularly at Bird Hill.

Holotype. — No. 26036, Walker Museum; *Paratype.* — No. 7886, University of Michigan.

Genus TENTACULITES Schlotheim

TENTACULITES FRAGILIS, n. sp.

(Plate XXXIII, Fig. 11)

Description. — Shell elongate-conical, round or slightly oval in section. Holotype, which is incomplete, has a length of 3.3 mm. and a maximum diameter of .5 mm. The apical portion is extremely small with very fine and regular annulations. Toward the anterior extremity the annulations become larger and somewhat irregular in size as well as in distance apart. A paratype, with a diameter of .9 mm., has 8 annulations in 1 mm. The large annulations are frequently marked by transverse striae.

Remarks. — This species is characterized by its small size and irregular annulations in the anterior portion of the shell.

Occurrence. — Throughout the Spirifer zone.

Holotype. — No. 8223; *Paratype.* — No. 8224, University of Michigan.

Class CEPHALOPODA

Genus ORTHOCERAS Breynius

ORTHOCERAS SP. 1

(Plate XXXVIII, Fig. 17)

Collections from all localities contain fragmental casts of *Orthoceratites*, probably *Orthoceras*, numbering from 1 to 6 segments. The diameter varies between 19 and 28 mm., and the distance from suture to suture averages about 6 mm. The shells tapered gradually, and probably reached lengths of 20 to 30 cm.

Occurrence. — Throughout the Spirifer zone, and sparingly in the Owen.

Figured specimen. — No. 7877, University of Michigan.

ORTHOCERAS SP. 2

(Plate XXXVIII, Figs. 1–3)

The singular *Orthoceratite* illustrated in these figures bears a certain resemblance to similarly preserved specimens of *Orthoceras duseri* of the Cincinnatian. The siphuncle is small and subcentral. The other characters are shown in the figures.

Occurrence. — Spirifer zone; rare.

Figured specimen. — No. 26000, Walker Museum.

GENUS GOMPHOCERAS SOWERBY

GOMPHOCERAS FLOYDENSIS, n. sp.

(Plate XXXVIII, Fig. 10)

Description. — Shell of moderate length, with the sections preserved as fossils commonly from 40 to 50 mm. in length. Segments average 3 to 4 mm. in length. In several specimens those segments nearest the living chamber are horizontally wrinkled, and from 2 to 3 mm. in length. In other specimens no such constriction is shown. Living chamber never fully preserved. In the holotype the preserved portion has a length of 21 mm., and is sharply constricted. In life, it probably possessed a shape much like that of the Silurian species *G. scrinium.* Surface of shell unknown.

Remarks. — This species of *Gomphoceras* is typical of the Spirifer zone, and is characterized by the large, sharply constricted living chamber.

Occurrence. — Middle and upper portions of the Spirifer zone, especially the Gastropod faunule at Rockford.

Holotype. — No. 26049; *Paratype.* — No. 26050, Walker Museum.

GOMPHOCERAS SP.

(Plate XXXVIII, Fig. 11)

Description. — This cephalopod cast, from the Stromatoporella faunule at Hackberry Grove, differs markedly from typical casts of *G. floydensis* in the very marked posterior constriction of the cone. It may, however, represent the posterior portion of a *floydensis* shell, but cannot be so identified with certainty.

Figured specimen. — No. 26051, Walker Museum.

Genus SPYROCERAS Hyatt

SPYROCERAS (?) FALSUM, n. sp.

(Plate XL, Fig. 17)

Description. — Shell large, apparently enlarging at a moderate rate, with the cross-section approximately circular. Internal characters unknown. Shell marked by broad, heavy, rounded rings, which in the type have intervals of 3 to 4 cm., the measurements being taken from crest to crest. In the paratype the distance is about 2.5 mm. Apparently the surface was marked by finer, sharper circular ridges, but there is no trace of longitudinal markings.

Remarks. — This species is known only from molds in stromatoporoid masses, found near Rockford, Iowa. So far as we are aware, three specimens have been found. The species may be distinguished from other Hackberry cephalopods by its notably larger size and strong annular ridges.

Occurrence. — The three specimens known have been found in the middle portion of the Spirifer zone, some feet below the Leptostrophia bed at the plant of the Rockford Brick and Tile Company.

Holotype. — No. 26052, Walker Museum; *Paratype.* — Collection of C. L. Webster.

Genus MANTICOCERAS Hyatt

MANTICOCERAS REGULARE, n. sp.

(Plate XXXIX, Figs. 1–3)

Manticoceras patersoni Clarke, Rept. N. Y. State Geol. for 1896, p. 45. 1907.
Manticoceras pattersoni C. L. Fenton, Am. Journ. Sci., 4th ser., vol. 48, p. 373. 1919.

Description. — Shell large, discoid; diameters of a representative individual 95 mm. and 110 mm.; thickness not determinable. Volutions 4 or 5, the outer ones embracing the inner to a depth of about one fourth the dorso-ventral diameter. Umbilicus broad and deep, exposing the inner volutions. Transverse section somewhat cuneiform, with the lateral surface almost flat and the curve at the periphery sharp. Living chamber of moderate size, occupying less than one third of the outer volution, and with dorso-ventral diameter (in the holotype) of about 45 mm. It appears to be slightly constricted at the aperture, but this condition probably is due to defective preservation. Air chambers numerous, those on the outer whorl number 4 in the space of 40 mm. Septa strong, somewhat thickened at the exterior margins. The characters of the septa are sufficiently well illustrated to make description unnecessary.

Remarks. — This species has been variously identified as *Manticoceras intumescens* and *M. patersoni*. From the latter species it differs in its flatter sides, more cuneiform cross-section, lesser rate of shell expansion, and much smaller body chamber. According to Hall's definition, the living chamber of *Manticoceras patersoni* is greater than the rest of the shell; in the Hackberry species it forms less than half of a single volution. The outline of *M. patersoni* given by Prosser and Swartz shows clearly that the two species are distinct.

Occurrence. — Found in the middle and upper portions of the Spirifer zone, mainly at Rockford and Hackberry Grove. Specimens generally small, fragmentary, and poorly preserved.

Types. — Collection of C. L. Webster.

GYROCERAS (?) SP.

(Plate XXXIX, Fig. 4)

In the Webster collection is a single mold, formed in the epitheca of a stromatoporoid, of a large cephalopod that may belong to the genus *Gyroceras*. The curvature is pronounced;

the annulations irregularly spaced, dichotomizing and heavy. The specimen is from the Owen substage.

Figured specimen. — Collection of C. L. Webster.

Phylum ANNELIDA (ANNULATA)

Class CHAETOPODA

Genus SERPULA Linne

SERPULA WELLERI, n. sp.

(Plate XLI, Fig. 5)

Description. — Tube 2 to 3 mm. in length and .4 to 1 mm. in diameter at the orifice. It is straight or slightly irregular, and marked by heavy wrinkles of growth. The orifice is circular, and directed upward at an angle of about 45 degrees. The worn shell presents a slightly pitted appearance, but definite tubuli are lacking.

Remarks. — This appears to be one of the least common annelids of the Hackberry. It also exhibits a definite preference for the shells of *Spirifers* as bases of attachment; almost every specimen in our collections is found upon these shells. It may be distinguished from other straight tubicolous annelids of the formation by its small, heavy, wrinkled tubes.

Occurrence. — Throughout the Spirifer zone of the Hackberry, as well as in the Gypidula faunule. Uncertain in the Owen.

Holotype. — No. 7865; *Paratype.* — No. 7866, University of Michigan.

SERPULA ANNULATA, n. sp.

(Plate XL, Figs. 10–11)

Description. — Tube 1 to 2 mm. in length, and .7 to 1 mm. in diameter at or near the aperture. Aperture circular or slightly subrectangular; it opens horizontally or at an angle of about 30 degrees to the plane of the host. Tube shows neither pits nor tabulae. It is marked by coarse growth wrinkles and heavy, angular annulae which persist even in weathered specimens.

Remarks. — This species differs from *S. welleri,* also of the Hackberry, in smaller size, more pronounced growth wrinkles, and the possession of the strong annulae. Of the latter there is no trace whatever in the larger species, the undulations of the shell being rounded in all cases. There is also a difference in host. *S. annulata* has been found only on small bryozoa, mainly of the genus *Orthopora,* while *S. welleri* generally grows on brachiopods, and particularly on *Spirifers.* It appears, also, that the tubes of *annulata* are much thinner than those of *welleri,* even when the individuals are of the same size.

Occurrence. — Upper portions of the Spirifer zone, especially the Strophonella faunule. Found only where bryozoa are abundant.

Holotype. — No. 7874; *Paratypes.* — Nos. 7873, 7875, University of Michigan, and 26075, Walker Museum.

Genus SPIRORBIS Lamarck

SPIRORBIS HACKBERRYENSIS, n. sp.

(Plate XLI, Fig. 6)

Spirorbis omphalodes C. L. Fenton, Am. Journ. Sci., 4th ser., vol. 48, pp. 369 and 371. 1919.

Description. — Tube closely coiled dextrally, with orifice somewhat expanded; volutions 2 to 3. Tube attached by one side of all the whorls; in some specimens the orifice is directed slightly upward. In such cases it is either circular or distorted and more or less crushed, as in the larger of the two cotypes. Whorls regularly rounded, heavy, and wrinkled. Weathered portions show numerous coarse pits.

Remarks. — This is the commonest annelid of the Hackberry, being found from the Gypidula faunule through the Spirifer zone, and in the Owen above. It may be distinguished by its regular, *Straparollus*-like coiling, rounded tube and moderate enrolling. It is abundant on brachiopods and gastropods, and not uncommon on corals, bryozoa, crinoid stems, and the epithecal surfaces of stromatoporoids. Probably it is most abundant on the shells of *Schizophoria iowaensis* and *Spirifer hungerfordi.*

Occurrence. — Throughout the Hackberry, above the coarse basal beds in which casts alone are found. Perhaps in the Cedar Valley beds as well.

Cotypes. — No. 7868, University of Michigan; *Paratypes.* — No. 26076, Walker Museum.

SPIRORBIS ANGULOSUS, n. sp.

(Plate XL, Figs. 12–13)

Description. — Shell small; diameter of coil 1.2 to 1.9 mm. in mature specimens; diameter of tube near aperture, about 6 or 7 mm. Tube coiled dextrally, with 2 to 3 whorls. Aperture generally somewhat constricted. Near it the tube becomes free from attachment and is directed more or less vertically, the degree varying with age. Weathered surface appears slightly pitted; unweathered surface marked by growth lines and heavy, angular annulae, which become almost spinose upon the median line of the upper surface.

Remarks. — This species may be distinguished from others by its very small size, heavy, angular annulae, and vertically directed outermost whorl. Like *S. annulata*, it grows upon small bryozoa, the diameter of the tube often exceeding that of the host. Among other features that serve to distinguish it from such forms as *Autodetus slocomi* is the small area of attachment which allows the tube to assume a nearly circular cross-section throughout its length.

Occurrence. — Upper portion of the Spirifer zone, and particularly the Strophonella faunule. Common on *Orthopora* and allied Bryozoa.

Holotype. — No. 7871; *Allotype.* — No. 7872, University of Michigan.

Genus AUTODETUS Lindström

AUTODETUS SLOCOMI, n. sp.

(Plate XLI, Figs. 7–8)

Spirorbis arkonensis, C. L. Fenton, Am. Journ. Sci., 4th ser., vol. 48, p. 371. 1919.

Description. — Tube closely coiled, and enrolled upon the inner whorls; whorls 2 to 3. Tube expands rapidly, and is subtriangular in cross-section near the middle of the outer whorl; it is attached on one side of all whorls. Diameter ranges from 1 to 2 mm., with the vertical diameter about one half the longest diameter of the coil. The figured cotype measures 1.2 × .6 mm. Whorls flattened on the outer slopes, nearly vertical on the inner; marked by pits and heavy growth wrinkles.

Remarks. — This species is distinguished by its cone-like appearance, semi-vertical orifice, and very large outer whorl. It is less common than *S. hackberryensis*, but appears to have about the same habits of growth. The particular specimen bearing the cotypes is a medium sized *Schizophoria*, the umbonal regions of which contain a large colony of this annelid, all of them a bit below the average size.

Occurrence. — Throughout the Spirifer zone of the Hackberry, and in the Gypidula faunule.

Cotypes. — No. 7867, University of Michigan.

Phylum ECHINODERMATA

Class ECHINOIDEA

Genus NORTONECHINUS Thomas

Nortonechinus C. L. Fenton, Am. Journ. Sci., 4th ser., vol. 48, p. 371. 1919. (*nomen nudum*)

Nortonechinus C. L. Fenton, Am. Mid. Nat., vol. 6, p. 191. 1920. (*nomen nudum*)

Nortonechinus Thomas, Bull. G. S. A., vol. 31, p. 212. 1920.

Nortonechinus Fenton and Fenton, Am. Mid. Nat., vol. 8, p. 219. 1923. (*Briefly defined*)

Nortonechinus Thomas, Echinoderms of the Lime Creek Beds of Iowa, p. 7, November, 1923. (*Full definition*)

Nortonechinus Thomas, Ia. Geol. Surv., vol. 29, p. 481, distributed Feb. 27, 1924. (*Identical with preceding*)

This echinoid, which probably resembled the modern Colobocentrotus,[26] has been carefully described and figured by Dr. A. O. Thomas of the University of Iowa in the references given. Figures are included here because the species *N. primus* is a very typical Hackberry fossil.

Remarks. — This genus is based on a single species; which is irregularly distributed throughout the upper half of the Spirifer zone. It is the commonest of the Hackberry echinoderms.

[26] See Lankester's *Treatise on Zoölogy,* part 3, *Echinoderma,* by F. A. Bather, pp. 313–314, Fig. 34.

NORTONECHINUS PRIMUS Fenton & Fenton

(Plate XL, Figs. 1–9; Plate XLI, Fig. 1)

Nortonechinus primus Fenton and Fenton, Am. Mid. Nat., vol. 8, p. 220,
figs. 1–8. 1923.
Nortonechinus welleri Thomas, Echinoderms of the Lime Creek Beds of Iowa.
November, 1923.
Nortonechinus welleri Thomas, Ia. Geol. Surv., vol. 29, p. 483, plates 47,
figs. 1–7; 48, 1–49; 49, 1–6 and 8–23. Distributed Feb. 27, 1924.

Description. — This species has been quite fully described
by Professor Thomas. The commonest spines are the broad,
flattened ones, which evidently covered the general surface of
the body. Other types are the elongate-polygonal ones, and the
heavy, club-shaped forms, which are marked by fine vertical
lining, and, in the latter case, are without granulose ornamenta-
tion on the distal surfaces. These two types of spine are uncommon, but may be found associated with the flattened one.
The third type is that shown in Plate 41, Fig. 1: an elongate
spine, heavier at the proximal end than at the distal. The
length of one of these, which is incomplete, is 16 mm.; diameter
at base 2.7 mm.; diameter about mid-way of the length, 2.8
mm. Probably these spines corresponded in position to the
long, flattened ones of *Colobocentrotus.*

Remarks. — Although the authors would prefer to withdraw
the name *primus* in favor of Dr. Thomas's *welleri*, in view of
his superior description, the course is impossible. Our brief
publication in the *American Midland Naturalist* appeared almost
exactly two months prior to the first publication by Dr. Thomas.
We do not, however, presume to add to his description except in
one minor point, which is to call attention to the fact that the
long spines shown in Plate XLI, Fig. 1, are considerably larger
than the spines figured by Dr. Thomas.

Occurrence. — Throughout the Spirifer zone, appearing near
the base of that division.

Holotype. — No. 8230; *Paratypes.* — Nos. 8231 to 8235, University of Michigan, Nos. 29109 and 26110, Walker Museum.

XENOCIDARIS AMERICANA Thomas

(Plate XLI, Fig. 2.)

Xenocidaris americana Thomas, Echinoderms of the Lime Creek Beds of Iowa, p. 13. November, 1923.
Xenocidaris americana Thomas, Ia. Geol. Surv., vol. 29, p. 50, figs. 1–25. Distributed Feb. 27, 1924.

A single specimen of this species is illustrated for comparison with figures of Nortonechinus.   A full description of the species will be found in Dr. Thomas's excellent paper on the echinoderms of the Iowa Devonian, referred to above.

Occurrence. — According to Dr. Thomas this species is typical of the lower Striatula zone.   We have found it rather common in the Lioclema faunule at Rockford.

Plesiotype. — No. 8236 University of Michigan.

PLATES AND EXPLANATIONS

NOTE

Unless otherwise stated, the figures are natural size; magnification or reduction is indicated in *diameters*. The following abbreviations indicate the locations of the specimens figured, catalogue numbers preceding the abbreviations:

U. M. = Geological Museum, University of Michigan;
W. M. = Walker Museum, University of Chicago;
C. L. W. = Collection of Clement L. Webster, Charles City, Iowa;
F. & F. = Collection of C. L. and M. A. Fenton;
U. S. N. M. = U. S. National Museum;
N. Y. S. M. = New York State Museum.

EXPLANATION OF PLATE I

PLATE I

1

4

8

2

5

10

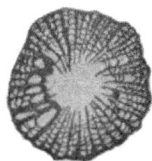

3

13

6

7

11

9

12

EXPLANATION OF PLATE II

PLATE II

1

2

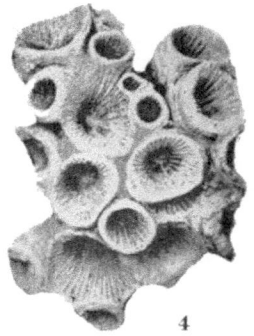

4

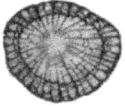

3

7

11

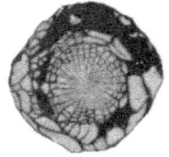

8

9

6

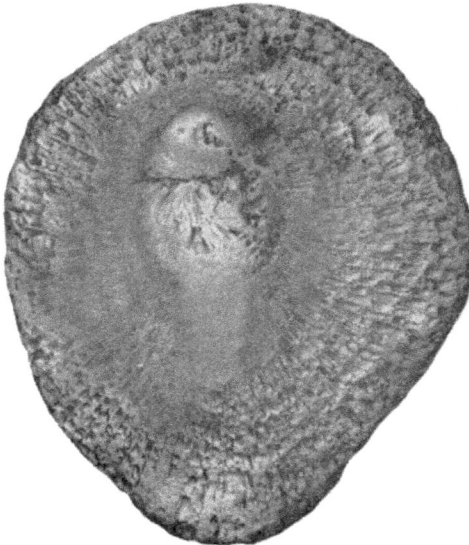

12

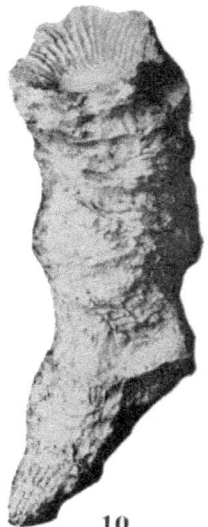

10

5

EXPLANATION OF PLATE III

PLATE III

3

2

4

9

5

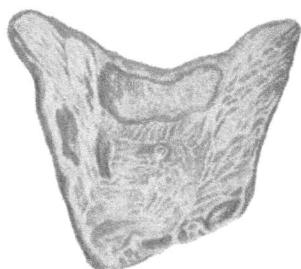

6

8

7

1

EXPLANATION OF PLATE IV

PLATE IV

1

2

3

4

5

EXPLANATION OF PLATE V

PLATE V

1

3

2

4

5

6

EXPLANATION OF PLATE VI

PLATE VI

3

1

2

10

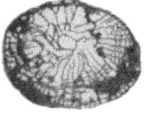

4

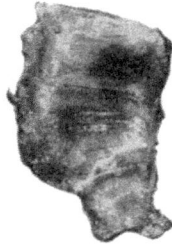

8

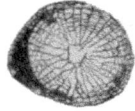

9

5

7

11

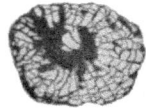

12

6

13

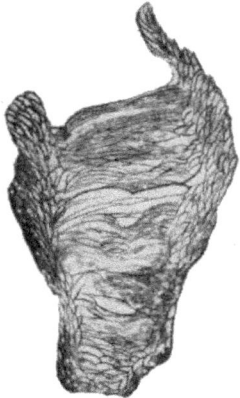

14

15

16

EXPLANATION OF PLATE VII

PLATE VII

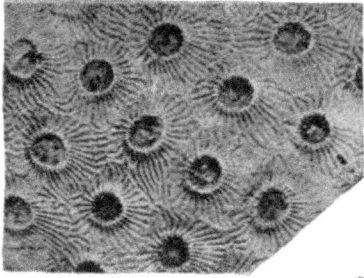

2

3

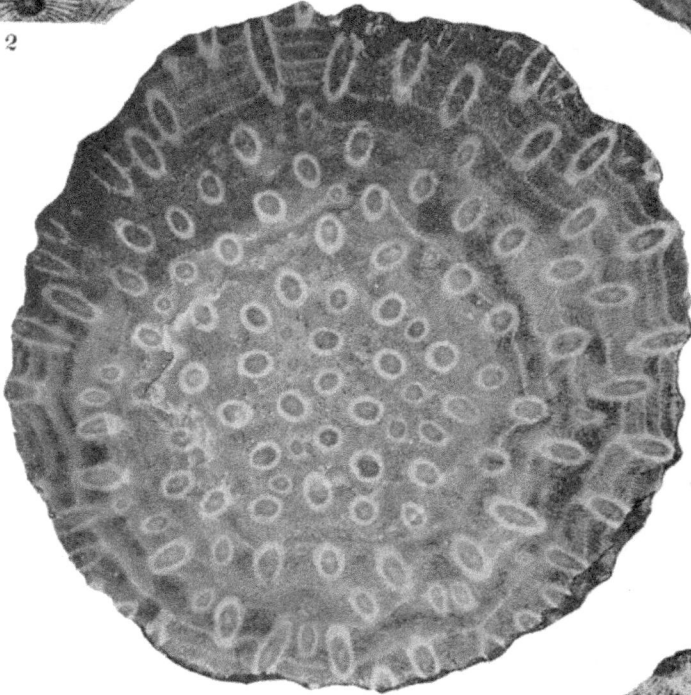

1

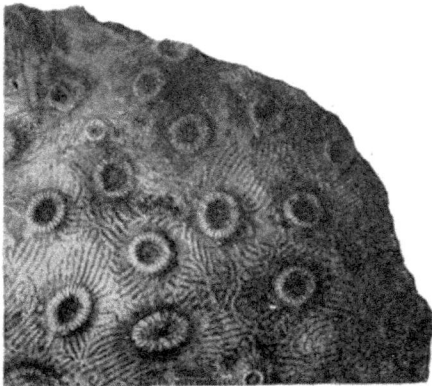

4

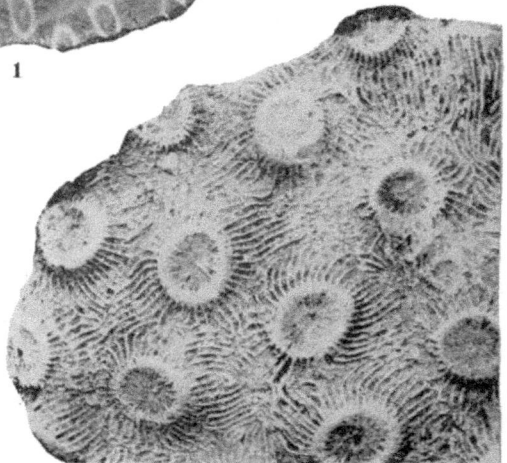

5

EXPLANATION OF PLATE VIII

PLATE VIII

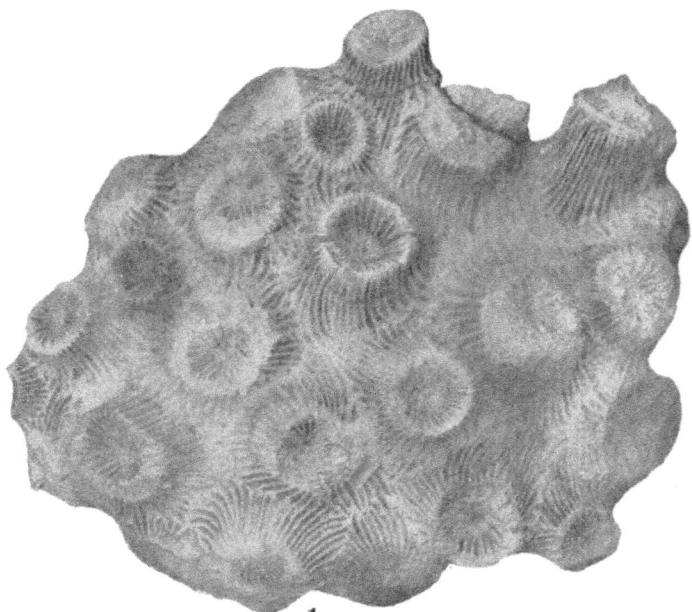

1

2

EXPLANATION OF PLATE IX

PLATE IX

1

6

2

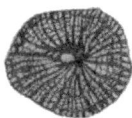

9

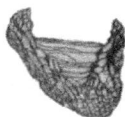

10

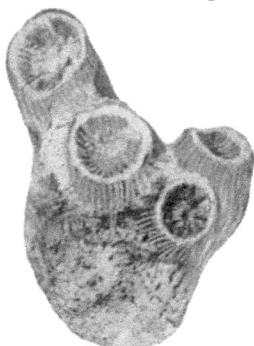

4

3

5

7

11

8

12

EXPLANATION OF PLATE X

PLATE X

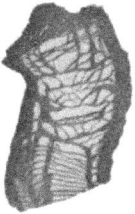

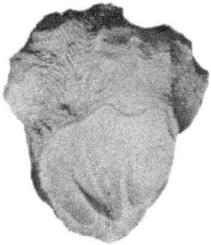

5

6

1

7

4

3

2

8

EXPLANATION OF PLATE XI

PLATE XI

2

1

EXPLANATION OF PLATE XII

PLATE XII

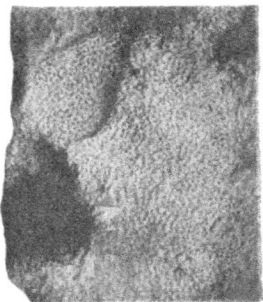

1

2

3

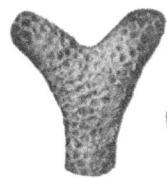

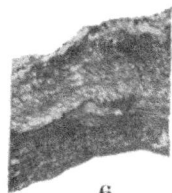

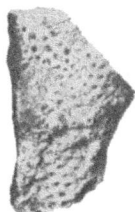

7

8

5

6

4

9

10

EXPLANATION OF PLATE XIII

PLATE XIII

1

2

EXPLANATION OF PLATE XIV

PLATE XIV

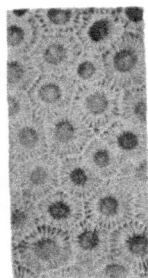

1

7

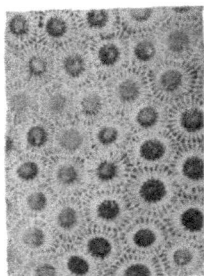

2

3

8

6

4

5

EXPLANATION OF PLATE XV

PLATE XV

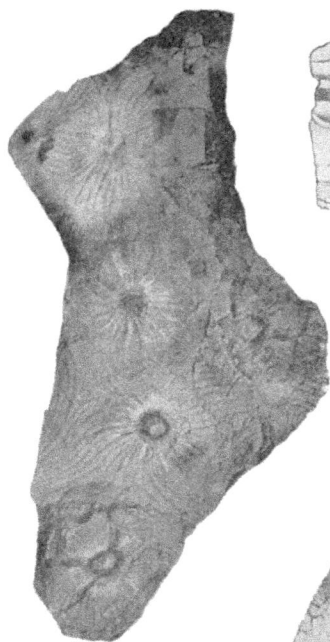

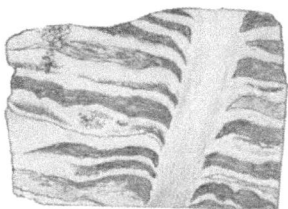

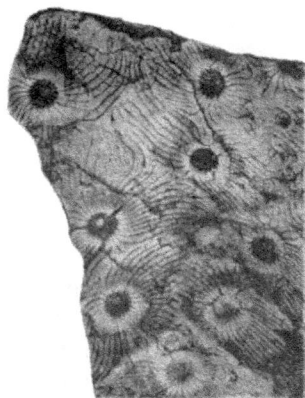

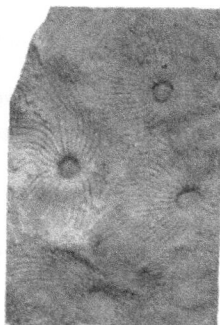

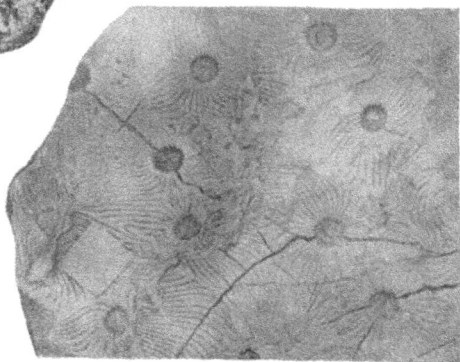

1

3

2

6

5

4

7

EXPLANATION OF PLATE XVI

PLATE XVI

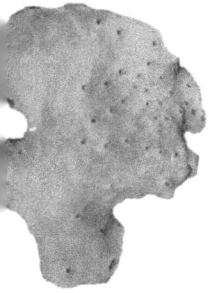

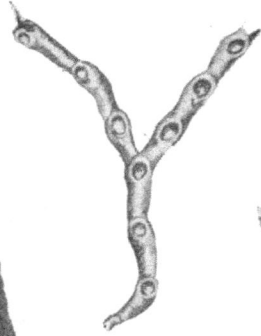

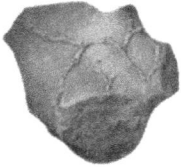

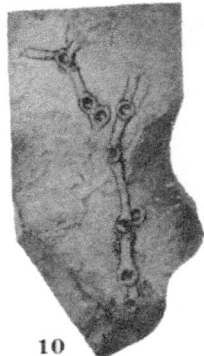

4

3

1

5

6

7

8

11

9

2

10

EXPLANATION OF PLATE XVII

PLATE XVII

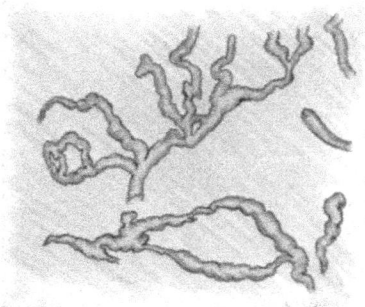

1

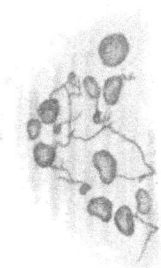

2

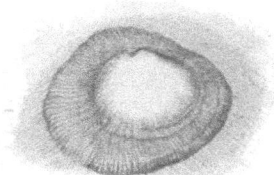

3

4

5

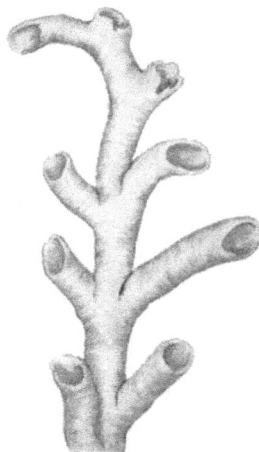

6

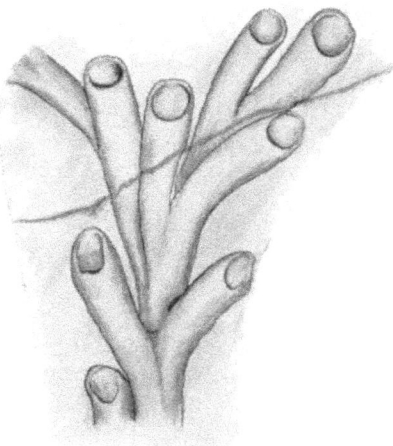

7

8

9

10

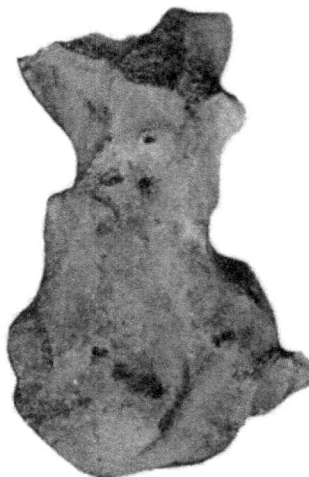

11

EXPLANATION OF PLATE XVIII

PLATE XVIII

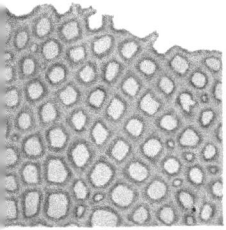

1

2

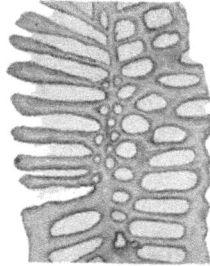

3

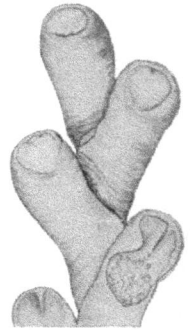

4

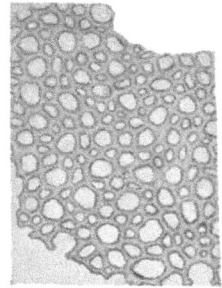

5

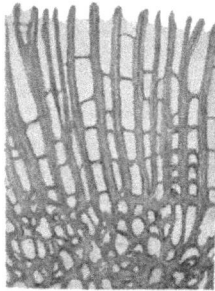

6

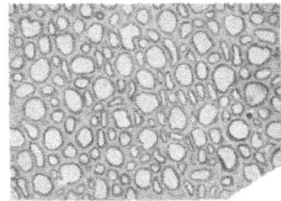

7

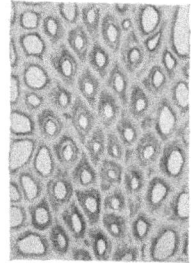

8

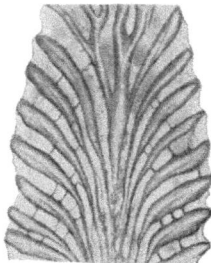

9

10

11

12

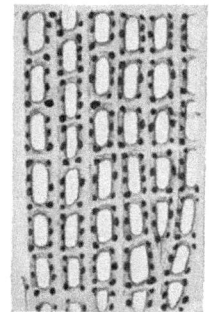

13

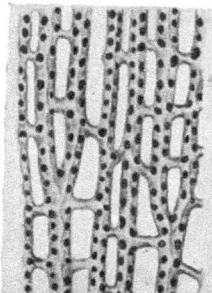

14

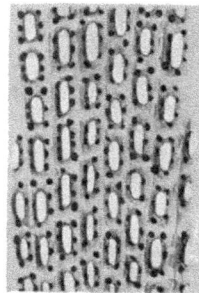

15

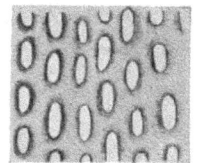

16

EXPLANATION OF PLATE XIX

PLATE XIX

1

2

4

3

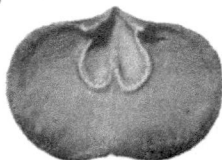

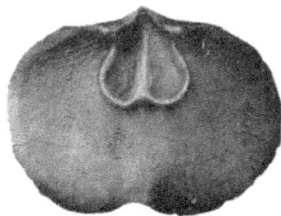

10

9

11

7

8

5

6

16

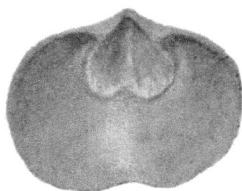

17

12

18

14

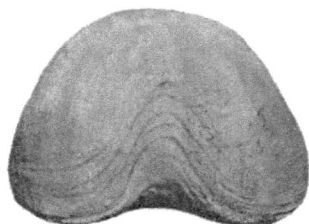

13

15

EXPLANATION OF PLATE XX

PLATE XX

1

2

3

4

5

6

7

8

9

11

14

12

15

10

13

16

17

18

19

20

26

27

28

21

22

24

25

23

EXPLANATION OF PLATE XXI

PLATE XXI

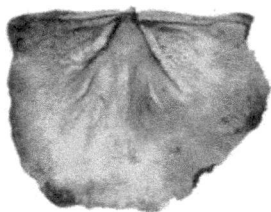

3

1

4

2

13

9

11

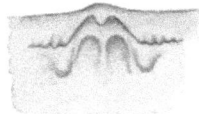

10

8

12

5

7

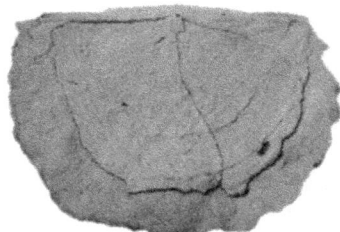

6

EXPLANATION OF PLATE XXII

[227]

PLATE XXII

1

3

5

7

14

2

4

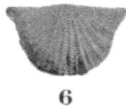

6

8

9

10

15

11

12

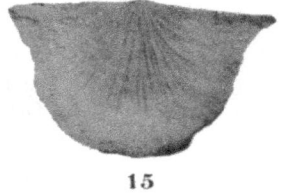

13

26

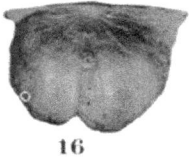

16

27

17

18

21

28

19

20

22

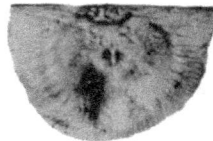

29

23

24

25

30

EXPLANATION OF PLATE XXIII

PLATE XXIII

1

2

3

4

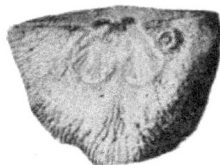

9

10

5

6

11

7

8

16

12

13

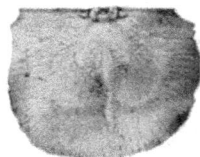

17

14

15

18

EXPLANATION OF PLATE XXIV

PLATE XVI

4

3

1

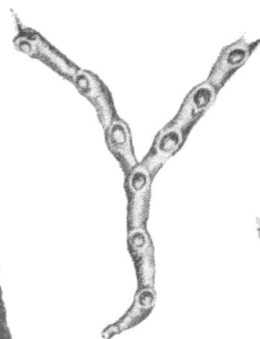

5

6

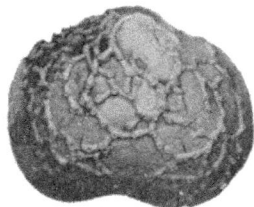

7

8

11

9

2

10

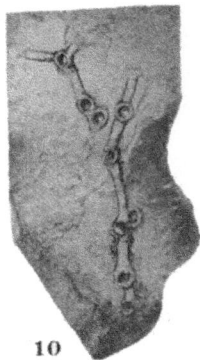

EXPLANATION OF PLATE XVII

PLATE XVII

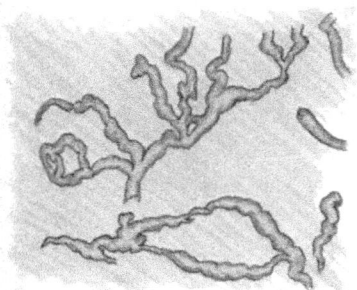

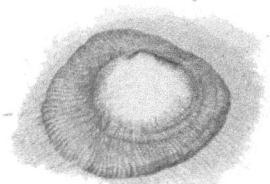

1

2

3

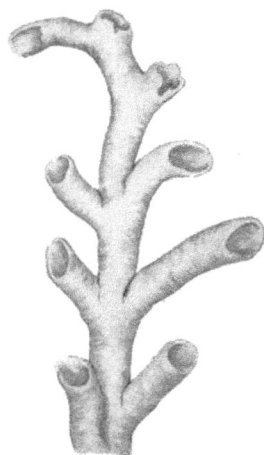

4

5

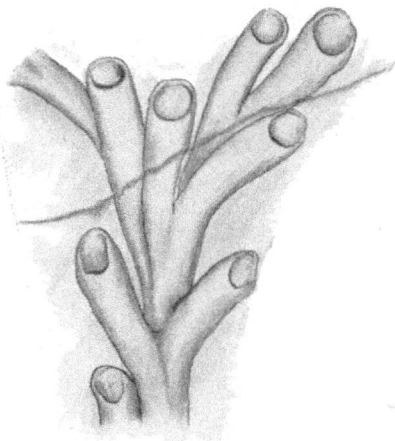

6

7

8

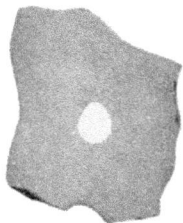

9

10

11

EXPLANATION OF PLATE XVIII

PLATE XVIII

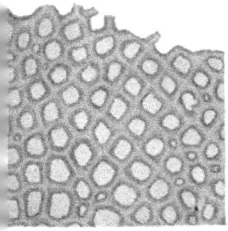

1

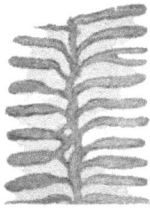

2

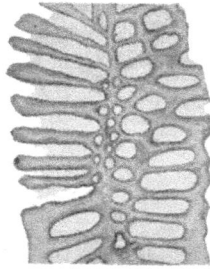

3

4

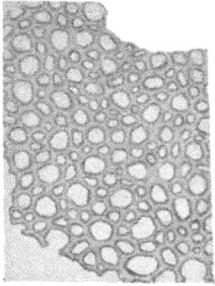

5

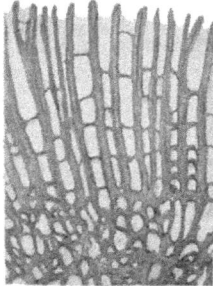

6

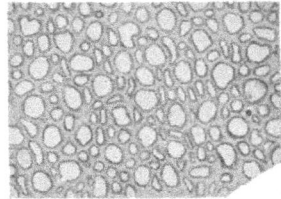

7

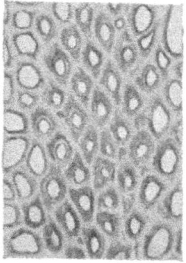

8

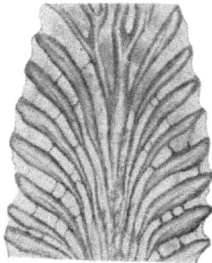

9

10

11

12

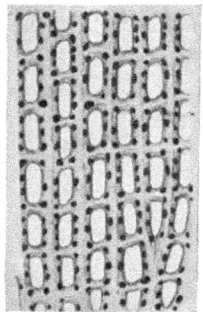

13

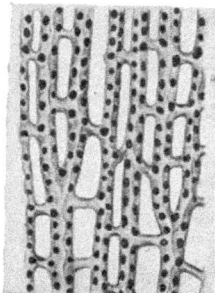

14

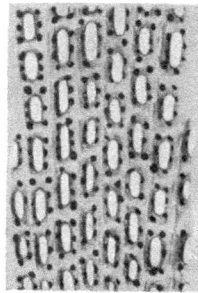

15

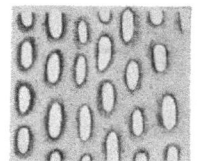

16

EXPLANATION OF PLATE XIX

PLATE XIX

1

2

4

3

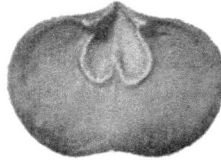

9

11

10

7

8

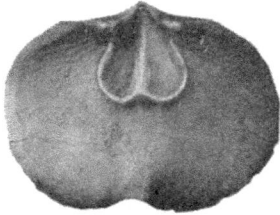

5

6

16

17

12

18

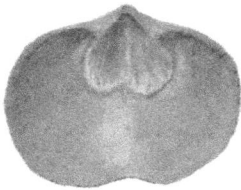

14

13

15

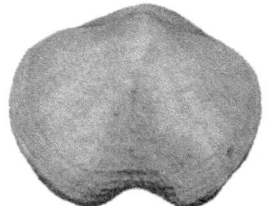

EXPLANATION OF PLATE XX

PLATE XX

1 2 3 4 5 6

7 8 9 11

14 12 15

10 13 16

17 18 19 20

26 27 28 21 22

24 25 23

EXPLANATION OF PLATE XXI

PLATE XXI

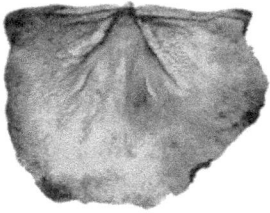

3

1

4

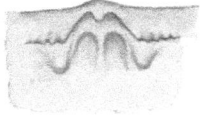

2

13

9

11

8

10

12

5

7

6

EXPLANATION OF PLATE XXII

[227]

PLATE XXII

1

3

5

7

14

2

4

6

8

15

9

10

11

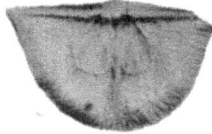
12

26

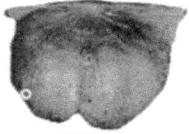
16

13

27

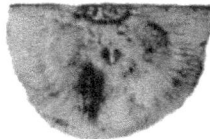
18

17

21

28

19

20

22

29

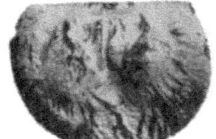
23

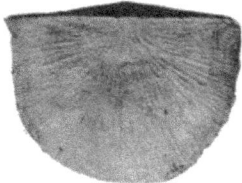
24

25

30

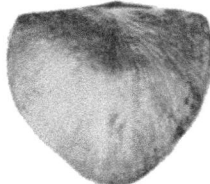

EXPLANATION OF PLATE XXIII

PLATE XXIII

1

2

3

4

9

10

5

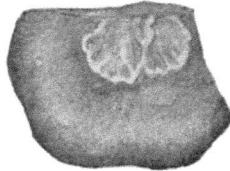

6

11

7

8

16

12

13

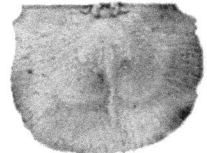

17

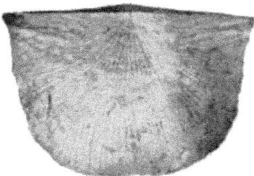

14

15

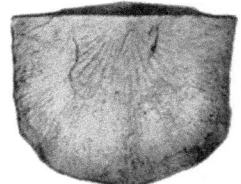

18

EXPLANATION OF PLATE XXIV

PLATE XXXII

1

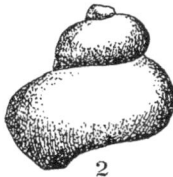

2

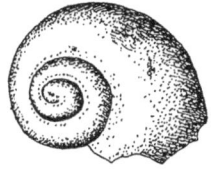

3

4

5

6

7

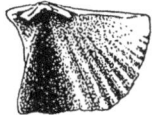

8

9

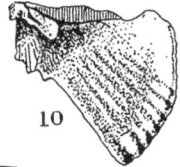

10

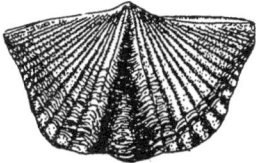

12

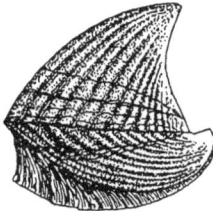

14

11

13

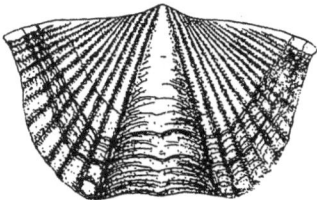

15

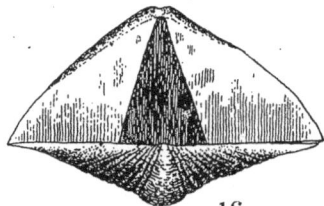

16

EXPLANATION OF PLATE XXXIII

PLATE XXXIII

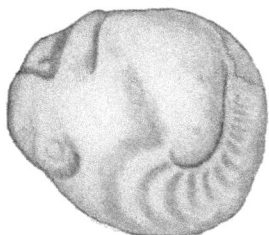

1

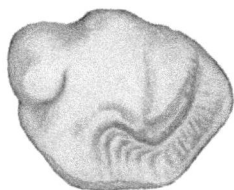

2

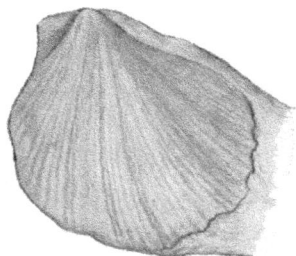

3

4

5

6

7

8

9

10

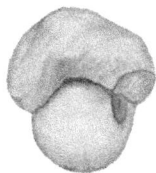

11

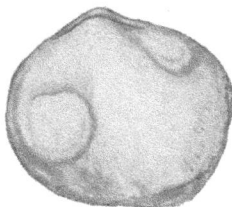

12

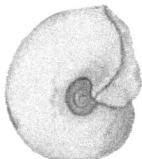

13

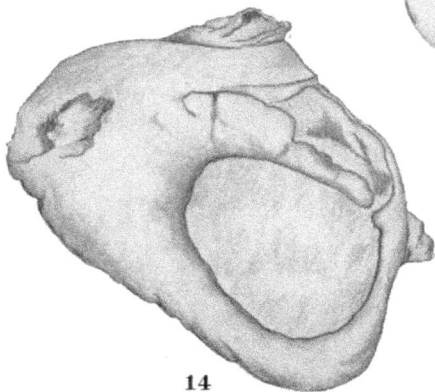

14

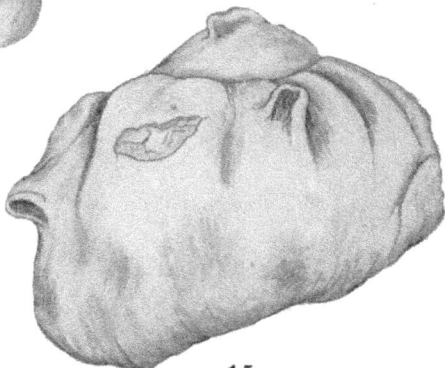

15

EXPLANATION OF PLATE XXXIV

PLATE XXXIV

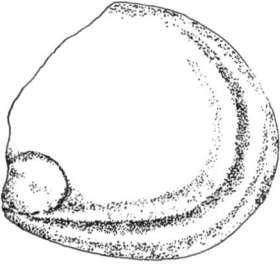

1

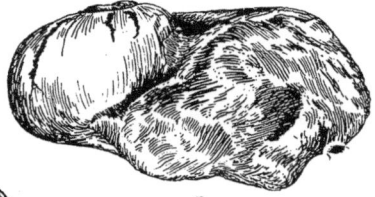

2

3

5

4

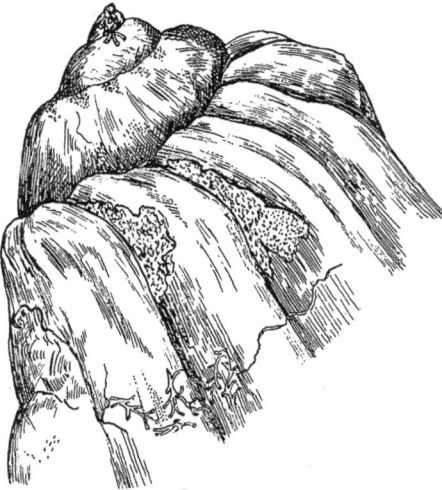

6

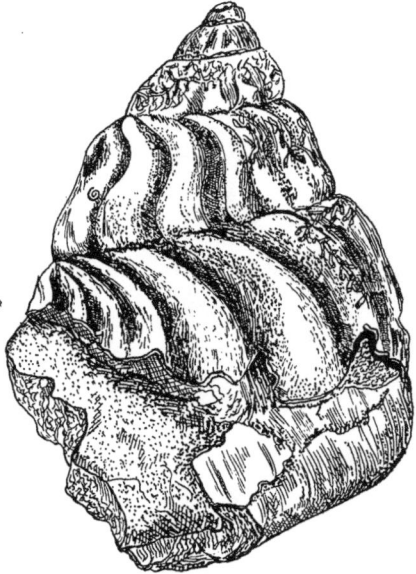

7

EXPLANATION OF PLATE XXXV

(All figures after Webster)

PLATE XXXV

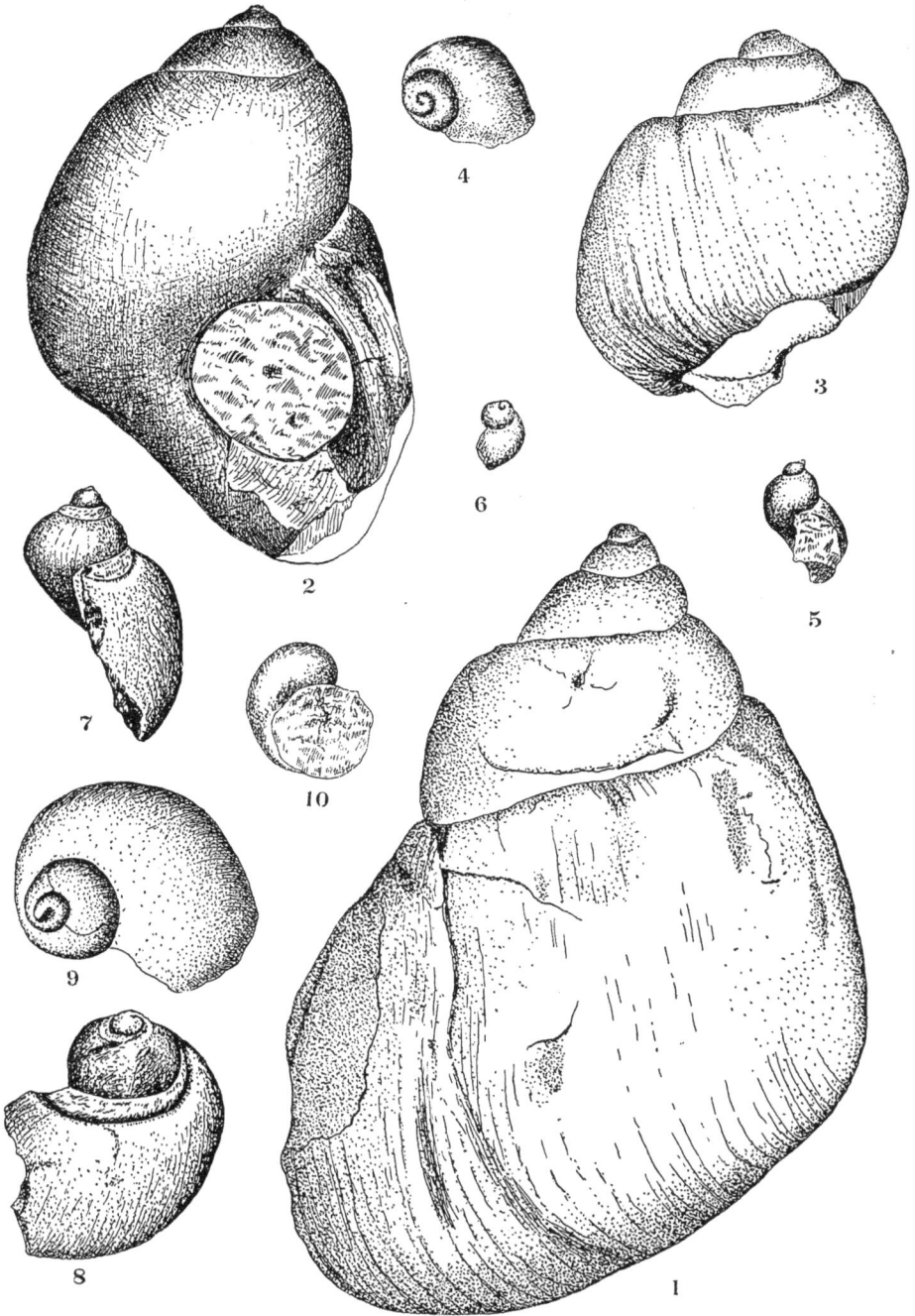

EXPLANATION OF PLATE XXXVI

(All figures after Webster)

PLATE XXXVI

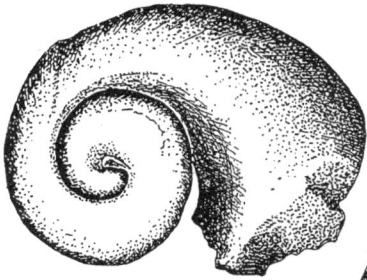

3

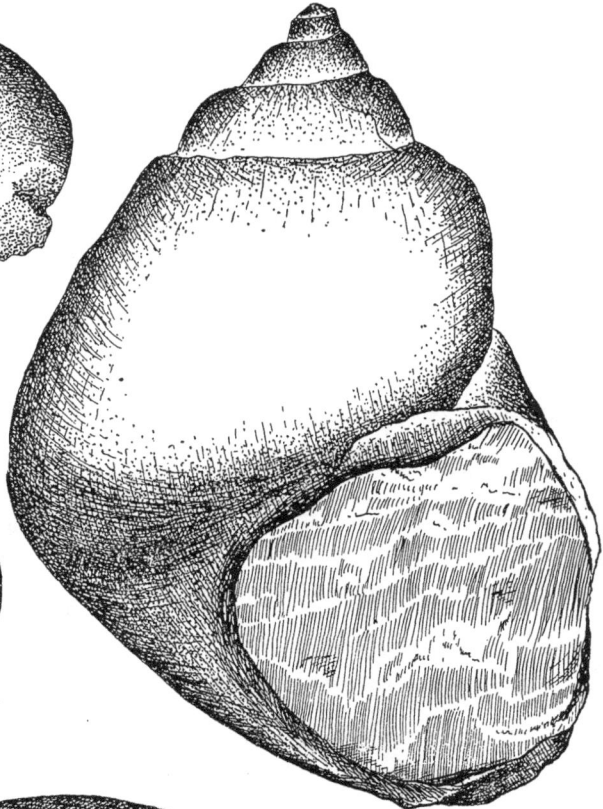

1

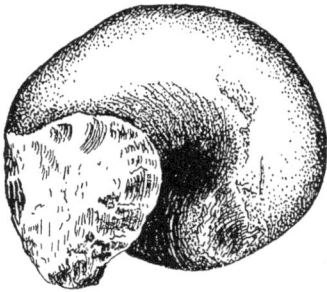

4

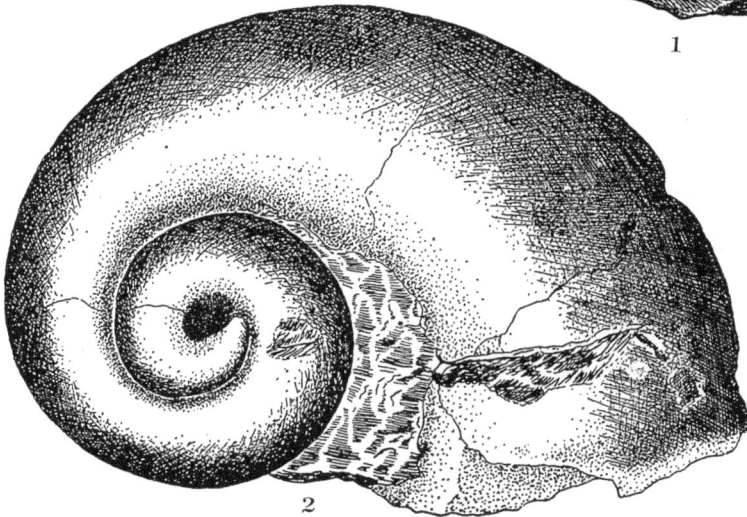

2

EXPLANATION OF PLATE XXXVII

(All figures after Webster)

PLATE XXXVII

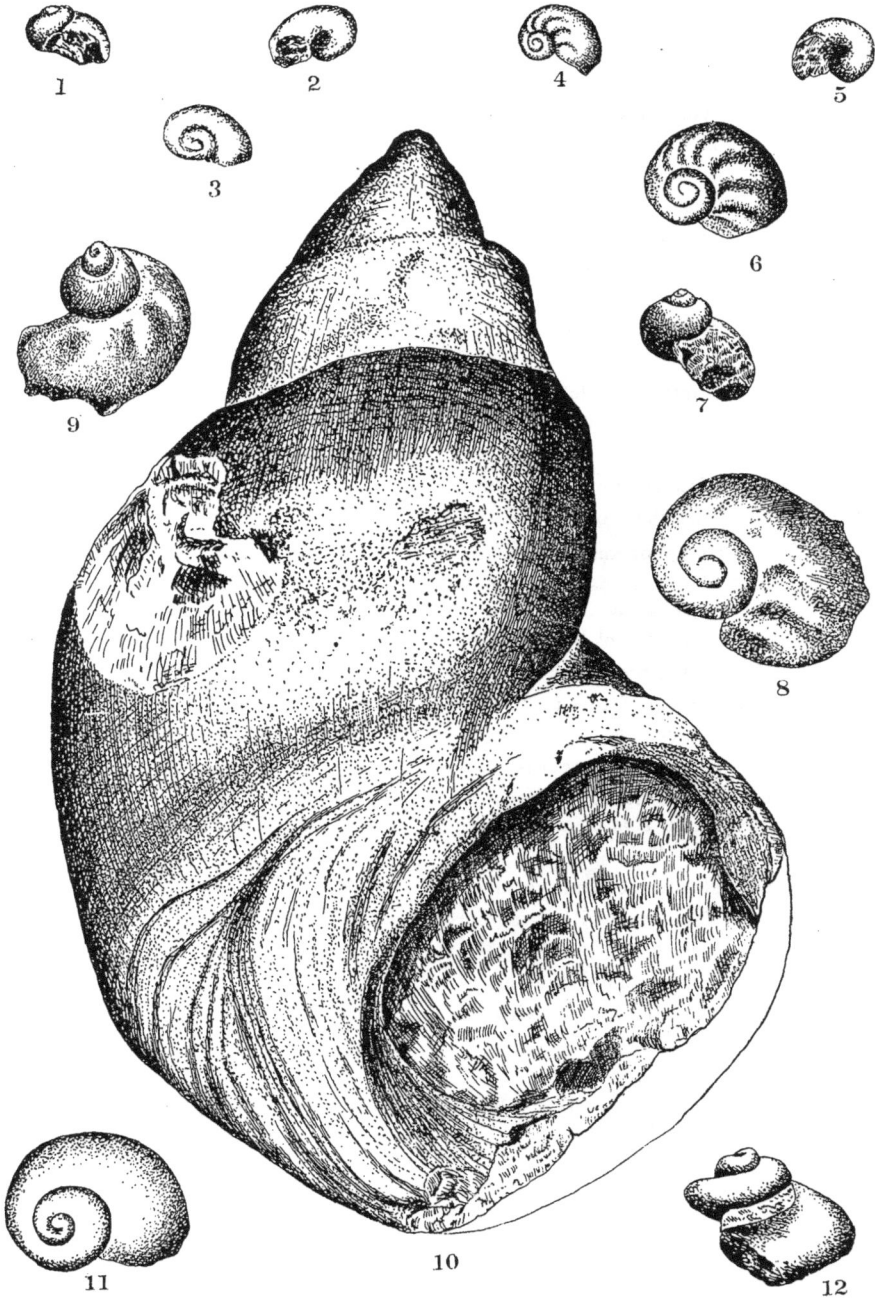

EXPLANATION OF PLATE XXXVIII

PLATE XXXVIII

1

2

3

4

5

6

7

8

9

10

11

12

13

14

15

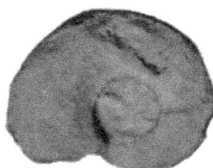

16

17

EXPLANATION OF PLATE XXXIX

PLATE XXXIX

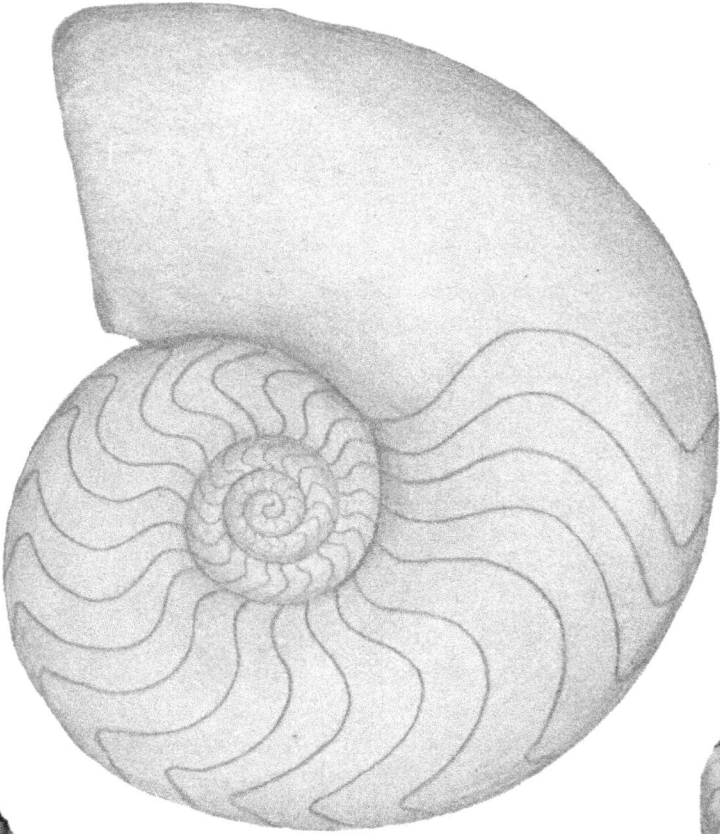

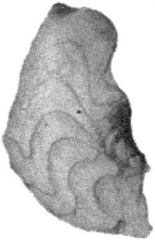

1

2

3

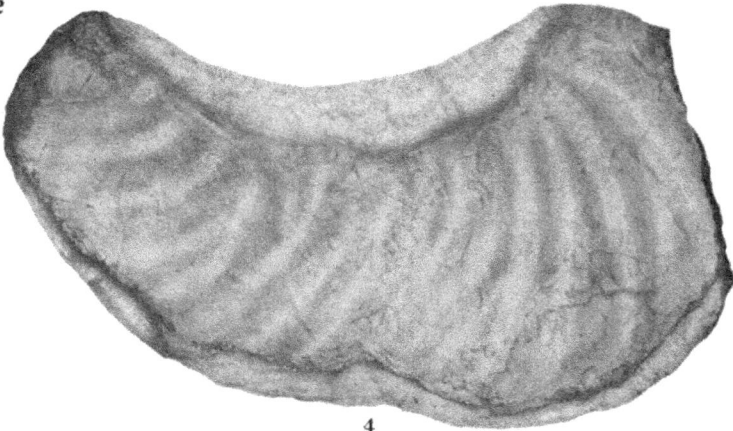

4

EXPLANATION OF PLATE XL

PLATE XL

1

2

3

4

5

7

8

9

10

11

12

13

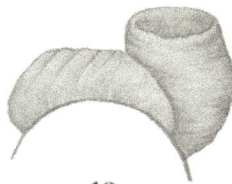

14

15

16

17

6

EXPLANATION OF PLATE XLI

PLATE XLI

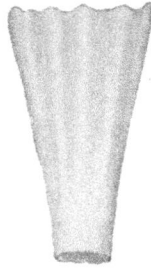

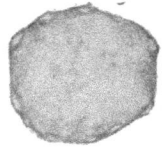

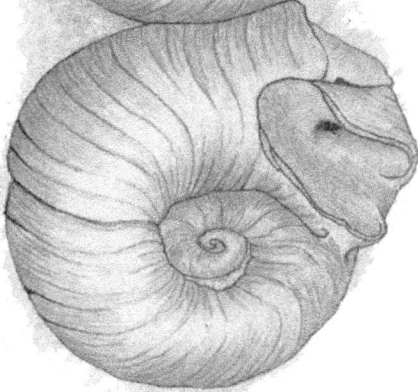

EXPLANATION OF PLATE XLII

*(A series of typical fossils from the Sheffield shale,
in the collection of C. L. Webster)*

FIGURE

1–2. Plants, undetermined, enlarged several diameters.

3–4. A fragment, probably of an arthropod; Fig. 4 is an enlargement
to show the surface markings.

5–8. Unidentified remains, probably borings of Annelids.

9–10. Two plant fragments, showing definite leaf form.

11. Fragment of what appears to be a minute cephalopod. × 5.

12. Lingula (?) fragilis Webster.

One of the cotypes. × 2.5.

PLATE XLII

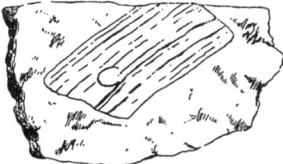

1

2

3

5

6

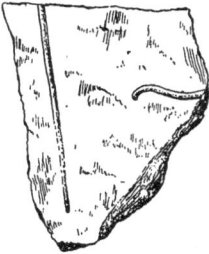

4

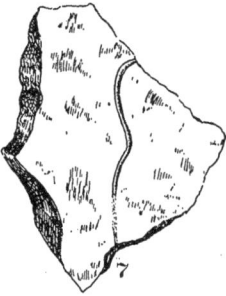

8

7

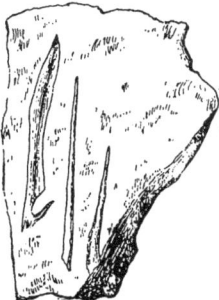

11

9

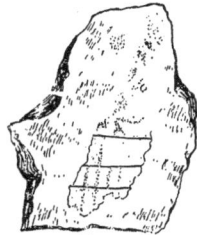

10

12

EXPLANATION OF PLATE XLIII

(All figures after Webster)

     1. Summit of a large cotype, showing the typical low spire (C. L. W.). The folds in the body whorl are partly obscured by compression.

  2–3. Summit and lateral views of another cotype, showing the strong folds (C. L. W.).

PLATE XLIII

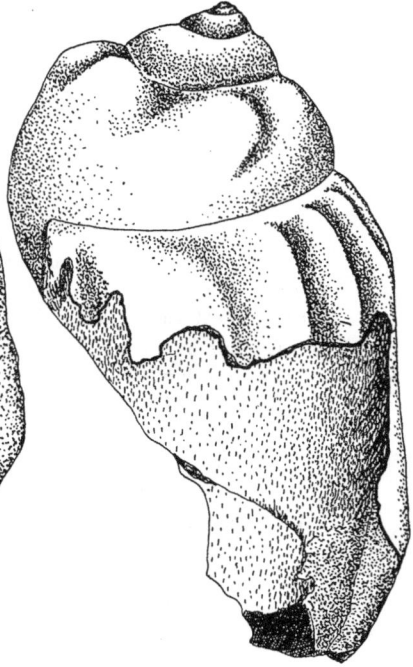

2

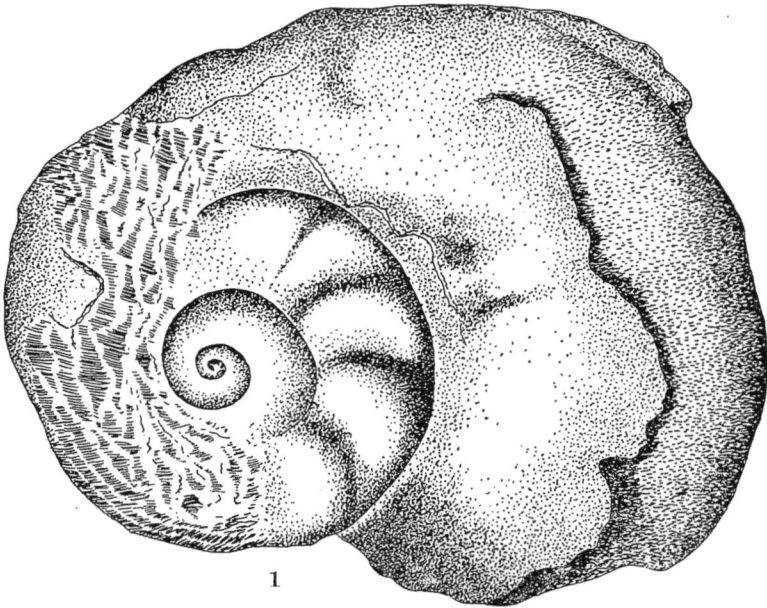

3

1

EXPLANATION OF PLATE XLIV

(All figures after Webster)

PLATE XLIV

4

5

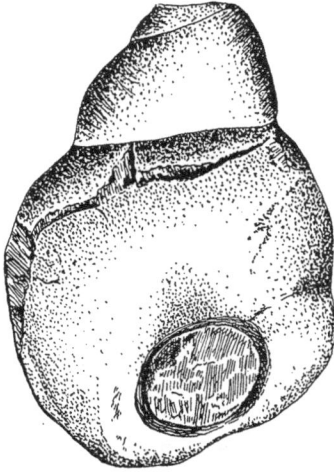

3

6

7

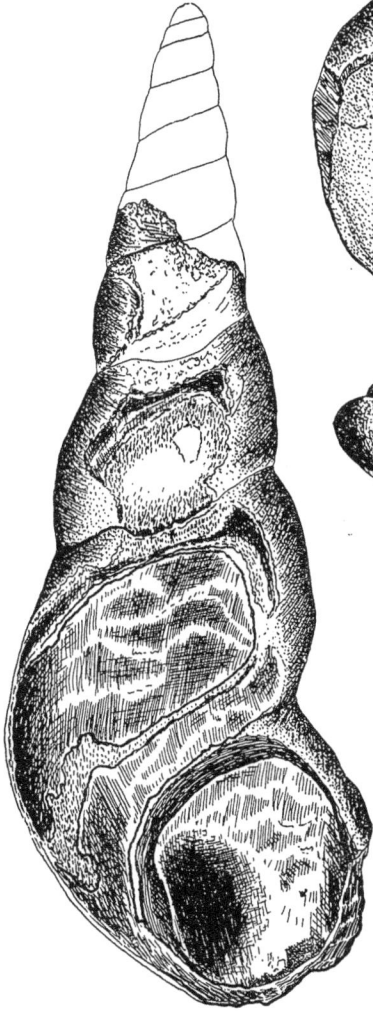

1

2

EXPLANATION OF PLATE XLV

(All figures after Webster)

PLATE XLV

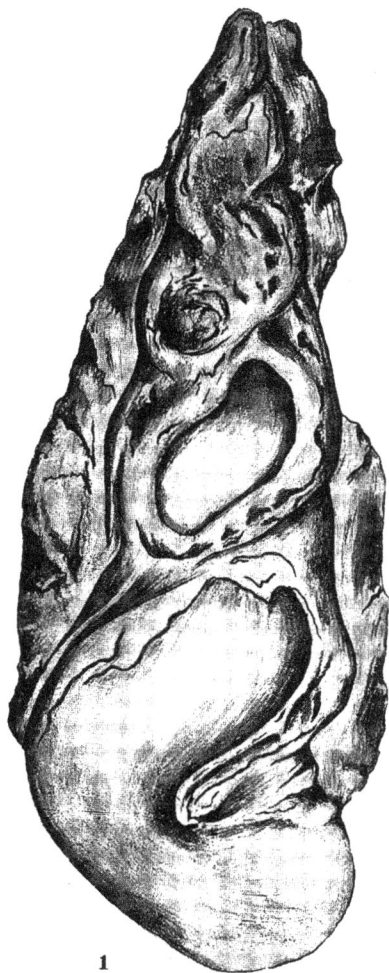

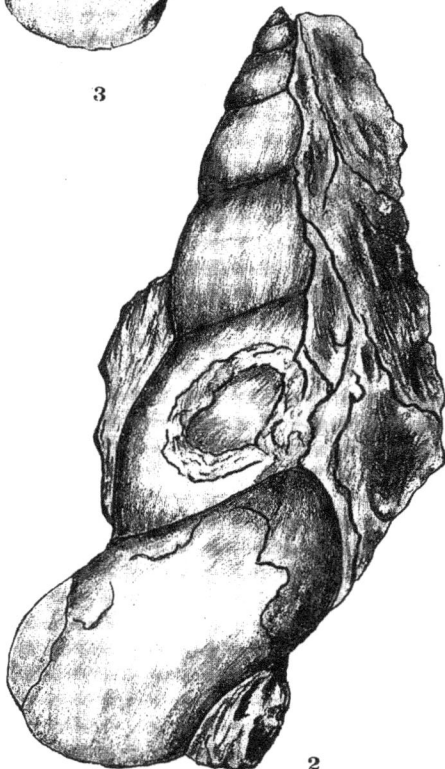

INDEX

Italics indicate synonyms; Arabic numerals refer to pages,
Roman numerals to plates.

UNIVERSITY OF MICHIGAN STUDIES

HUMANISTIC SERIES

General Editors: FRANCIS W. KELSEY AND HENRY A. SANDERS.

Size, 22.7 × 15.2 cm. 8°. Bound in Cloth.

Vol. I. ROMAN HISTORICAL SOURCES AND INSTITUTIONS. Edited by Henry A. Sanders, University of Michigan. Pp. vii + 402. $2.50 net.

CONTENTS

THE MACMILLAN COMPANY

Publishers   64–66 Fifth Avenue   New York

Vol. V. Sources of the Synoptic Gospels. By Rev. Dr. Carl S. Patton, First Congregational Church, Los Angeles, California. Pp. xiii + 263. $1.30 net.

Size, 28 × 18.5 cm. 4to.

Vol. VI. Athenian Lekythoi with Outline Drawing in Glaze Varnish on a White Ground. By Arthur Fairbanks, Director of the Museum of Fine Arts, Boston. With 15 plates, and 57 illustrations in the text. Pp. viii + 371. Bound in cloth. $4.00 net.

Vol. VII. Athenian Lekythoi with Outline Drawing in Matt Color on a White Ground, and an Appendix: Additional Lekythoi with Outline Drawing in Glaze Varnish on a White Ground. By Arthur Fairbanks. With 41 plates. Pp. x + 275. Bound in cloth. $3.50 net.

Vol. VIII. The Old Testament Manuscripts in the Freer Collection. By Henry A. Sanders, University of Michigan. With 9 plates showing pages of the Manuscripts in facsimile. Pp. viii + 357. Bound in cloth. $3.50 net.

Parts Sold Separately in Paper Covers:

Part I. The Washington Manuscript of Deuteronomy and Joshua. With 3 folding plates. Pp. vi + 104. $1.25 net.

Part II. The Washington Manuscript of the Psalms. With 1 single plate and 5 folding plates. Pp. viii + 105–349. $2.00 net.

Vol. IX. The New Testament Manuscripts in the Freer Collection. By Henry A. Sanders, University of Michigan. With 8 plates showing pages of the Manuscripts in facsimile. Pp. x + 323. Bound in cloth. $3.50.

Parts Sold Separately in Paper Covers:

Part I. The Washington Manuscript of the Four Gospels. With 5 plates. Pp. vii + 247. $2.00 net.

Part II. The Washington Manuscript of the Epistles of Paul. With 3 plates. Pp. ix + 251–315. $1.25 net.

Vol. X. The Coptic Manuscripts in the Freer Collection. By William H. Worrell, Hartford Seminary Foundation. With 12 plates. Pp. xxvi + 396. Bound in cloth. $4.75 net.

Parts Sold Separately in Paper Covers:

Part I. The Coptic Psalter. The Coptic text in the Sahidic Dialect, with an Introduction, and with 6 plates showing pages of the Manuscript and Fragments in facsimile. Pp. xxvi + 112. $2.00 net.

Part II. A Homily on the Archangel Gabriel by Celestinus, Bishop of Rome, and a Homily on the Virgin by Theophilus, Archbishop of Alexandria, from Manuscript Fragments in the Freer Collection and the British Museum. The Coptic Text with an Introduction and Translation, and with 6 plates showing pages of the Manuscripts in facsimile. Pp. 113–396. $2.50 net.

THE MACMILLAN COMPANY

Publishers   64–66 Fifth Avenue   New York

VOL. XI. CONTRIBUTIONS TO THE HISTORY OF SCIENCE.

Part I. ROBERT OF CHESTER'S LATIN TRANSLATION OF THE ALGEBRA OF AL-KHOWARIZMI. With an Introduction, Critical Notes, and an English Version. By Louis C. Karpinski, University of Michigan. With 4 plates showing pages of manuscripts in facsimile, and 25 diagrams in the text. Pp. vii + 164. Paper covers. $2.00 net.

Part II. THE PRODROMUS OF NICOLAUS STENO'S LATIN DISSERTATION CONCERNING A SOLID BODY ENCLOSED BY PROCESS OF NATURE WITHIN A SOLID. Translated into English by John G. Winter, University of Michigan, with a Foreword by Professor William H. Hobbs. With 7 plates. Pp. vii + 169–283. Paper covers. $1.30 net.

Part III. VESUVIUS IN ANTIQUITY. Passages of Ancient Authors, with a Translation and Elucidations. By Francis W. Kelsey. Illustrated. (*In preparation.*)

VOL. XII. STUDIES IN EAST CHRISTIAN AND ROMAN ART. By Charles R. Morey, Princeton University, and Walter Dennison. With 67 plates (10 colored) and 91 illustrations in the text. Pp. xiii + 175. $4.75 net.

Parts Sold Separately:

Part I. EAST CHRISTIAN PAINTINGS IN THE FREER COLLECTION. By Charles R. Morey. With 13 plates (10 colored) and 34 illustrations in the text. Pp. xiii + 86. Bound in cloth. $2.50 net.

Part II. A GOLD TREASURE OF THE LATE ROMAN PERIOD. By Walter Dennison. With 54 plates and 57 illustrations in the text. Pp. 89–175. Bound in cloth. $2.50 net.

VOL. XIII. DOCUMENTS FROM THE CAIRO GENIZAH IN THE FREER COLLECTION. Text, with Translation and an Introduction by Richard Gottheil, Columbia University. (*In press.*)

VOL. XIV. TWO STUDIES IN LATER ROMAN AND BYZANTINE ADMINISTRATION. By Arthur E. R. Boak and James E. Dunlap, University of Michigan. Pp. x + 324. Bound in cloth. $2.25 net.

Parts Sold Separately in Paper Covers:

Part I. THE MASTER OF THE OFFICES IN THE LATER ROMAN AND BYZANTINE EMPIRES. By Arthur E. R. Boak. Pp. x + 160. Paper covers. $1.00 net.

Part II. THE OFFICE OF THE GRAND CHAMBERLAIN IN THE LATER ROMAN AND BYZANTINE EMPIRES. By James E. Dunlap. Pp. 165–324. $1.00 net.

VOL. XV. GREEK THEMES IN MODERN MUSICAL SETTINGS. By Albert A. Stanley, University of Michigan. With 10 plates. Pp. xxii + 385. Bound in cloth. $4.00 net.

Parts Sold Separately in Paper Covers:

Part I. INCIDENTAL MUSIC TO PERCY MACKAYE'S DRAMA OF SAPPHO AND PHAON. Pp. 1–68. $0.90 net.

THE MACMILLAN COMPANY

Publishers   64–66 Fifth Avenue   New York

Part II. MUSIC TO THE ALCESTIS OF EURIPIDES WITH ENGLISH TEXT. Pp. 71–120. $0.80 net.

Part III. MUSIC TO THE IPHIGENIA AMONG THE TAURIANS BY EURIPIDES, WITH GREEK TEXT. Pp. 123–190. $0.75 net.

Part IV. TWO FRAGMENTS OF ANCIENT GREEK MUSIC. Pp. 217–225. $0.30 net.

Part V. MUSIC TO CANTICA OF THE MENAECHMI OF PLAUTUS. Pp. 229–263. $0.60 net.

Part VI. ATTIS: A SYMPHONIC POEM. Pp. 265–383. $1.00 net.

VOL. XVI. NICOMACHUS OF GERASA: INTRODUCTION TO ARITHMETIC. Translated into English by Martin Luther D'Ooge, with Studies in Greek Arithmetic by Frank Egleston Robbins and Louis C. Karpinski. (*In press.*)

VOLS. XVII, XVIII, XIX, XX. ROYAL CORRESPONDENCE OF THE ASSYRIAN EMPIRE. Translated into English, with a transliteration of the text and a Commentary. By Leroy Waterman, University of Michigan. (*In press.*)

VOL. XXI. THE PAPYRUS MINOR PROPHETS IN THE FREER COLLECTION AND THE BERLIN FRAGMENT OF GENESIS. By Henry A. Sanders, University of Michigan, and Carl Schmidt, University of Berlin. (*In press.*)

FACSIMILES OF MANUSCRIPTS

Size, 40.5 × 35 cm.

FACSIMILE OF THE WASHINGTON MANUSCRIPT OF DEUTERONOMY AND JOSHUA IN THE FREER COLLECTION. With an Introduction by Henry A. Sanders. Pp. x; 201 heliotype plates. The University of Michigan. Ann Arbor, Michigan, 1910.

Limited edition, distributed only to Libraries, under certain conditions. A list of Libraries containing this Facsimile is printed in *University of Michigan Studies, Humanistic Series*, Volume VIII, pp. 351–353.

Size, 34 × 26 cm.

FACSIMILE OF THE WASHINGTON MANUSCRIPT OF THE FOUR GOSPELS IN THE FREER COLLECTION. With an Introduction by Henry A. Sanders. Pp. x; 372 heliotype plates and 2 colored plates. The University of Michigan. Ann Arbor, Michigan, 1912.

Limited edition, distributed only to Libraries, under certain conditions. A list of Libraries containing this Facsimile is printed in *University of Michigan Studies, Humanistic Series*, Volume IX, pp. 317–320.

SCIENTIFIC SERIES

Size, 28 × 18.5 cm. 4°. Bound in Cloth.

VOL. I. THE CIRCULATION AND SLEEP. By John F. Shepard, University of Michigan. Pp. ix + 83, with an Atlas of 63 plates, bound separately. Text and Atlas, $2.50 net.

VOL. II. STUDIES ON DIVERGENT SERIES AND SUMMABILITY. By Walter B. Ford, University of Michigan. Pp. xi + 194. $2.50.

THE MACMILLAN COMPANY

Publishers   64–66 Fifth Avenue   New York

UNIVERSITY OF MICHIGAN PUBLICATIONS

HUMANISTIC PAPERS

General Editor: EUGENE S. McCARTNEY

Size, 22.7 × 15.2 cm. 8°.   Bound in Cloth.

THE LIFE AND WORKS OF GEORGE SYLVESTER MORRIS. A CHAPTER IN THE HISTORY OF AMERICAN THOUGHT IN THE NINETEENTH CENTURY. By ROBERT M. WENLEY, University of Michigan. Pp. xv + 332. Cloth. $1.50 net.

LATIN AND GREEK IN AMERICAN EDUCATION, WITH SYMPOSIA ON THE VALUE OF HUMANISTIC STUDIES. Edited by FRANCIS W. KELSEY. Pp. x + 396. $1.50.

> THE PRESENT POSITION OF LATIN AND GREEK, The Value of Latin and Greek as Educational Instruments, the Nature of Culture Studies.
>
> SYMPOSIA ON THE VALUE OF HUMANISTIC, Particularly Classical, Studies as a Preparation for the Study of Medicine, Engineering, Law and Theology.
>
> A SYMPOSIUM ON THE VALUE OF HUMANISTIC, Particularly Classical, Studies as a Training for Men of Affairs.
>
> A SYMPOSIUM ON THE CLASSICS AND THE NEW EDUCATION.
>
> A SYMPOSIUM ON THE DOCTRINE OF FORMAL DISCIPLINE IN THE LIGHT OF CONTEMPORARY PSYCHOLOGY.
> (*Out of print; new edition in preparation.*)

THE MENAECHMI OF PLAUTUS. The Latin Text, with a Translation by JOSEPH H. DRAKE, University of Michigan. Pp. xi + 130. Paper covers. $0.60.

THE SENATE AND TREATIES, 1789–1817: THE DEVELOPMENT OF THE TREATY-MAKING FUNCTIONS OF THE UNITED STATES SENATE DURING THEIR FORMATIVE PERIOD. By RALSTON HAYDEN, University of Michigan. Pp. xvi + 237. Cloth. $1.50 net.

Size, 23.5 × 15.5 cm. 8°.   Bound in Cloth.

WILLIAM PLUMER'S MEMORANDUM OF PROCEEDINGS IN THE UNITED STATES SENATE, 1803–1807. Edited by EVERETT SOMERVILLE BROWN, University of Michigan. Pp. xi + 673. Cloth. $3.50.

PAPERS OF THE MICHIGAN ACADEMY OF SCIENCE, ARTS AND LETTERS

(containing Papers submitted at Annual Meetings)

Editors: PAUL S. WELCH and EUGENE S. McCARTNEY

Size, 24.2 × 16.5 cm. 8°. Bound in Cloth.

VOL. I (1921). With 38 plates and 5 maps. Pp. xi + 424. $2.00 net.

VOL. II (1922). With 11 plates. Pp. xi + 226. $2.00 net. Bound in paper $1.50 net.

VOL. III (1923). With 26 plates, 15 text figures and three maps. Pp. xiii + 473. $3.00 net. Bound in paper, $2.25 net.

THE MACMILLAN COMPANY

Publishers   64–66 Fifth Avenue   New York

CONTRIBUTIONS FROM THE MUSEUM OF GEOLOGY

VOLUME I

THE STRATIGRAPHY AND FAUNA OF THE HACKBERRY STAGE OF THE UPPER DEVONIAN. By Carroll Lane Fenton and Mildred Adams Fenton. Pp. xi + 260. Cloth. $2.75.

VOLUME II

(*All communications relative to the Numbers of Volume II should be addressed to the Librarian, General Library, University of Michigan.*)

No. 1. A Possible Explanation of Fenestration in the Primitive Reptilian Skull, with Notes on the Temporal Region of the Genus Dimetrodon, by E. C. Case. Pp. 1–12, with five illustrations. $0.30.

No. 2. Occurrence of the Collingwood Formation in Michigan, by R. Ruedemann and G. M. Ehlers. Pp. 13–18. $0.15.

No. 3. Silurian Cephalopods of Northern Michigan, by Aug. F. Foerste. Pp. 19–104, with 17 plates. $1.00.

_____

(*All communications relative to the volumes listed below should be addressed to the Librarian, General Library, University of Michigan.*)

HISTORICAL STUDIES

(Published under the direction of the Department of History, 1911–1913.)

VOL. I. A HISTORY OF THE PRESIDENT'S CABINET. By Mary Louise Hinsdale. Pp. ix + 355. Cloth. $2.00.

VOL. II. ENGLISH RULE IN GASCONY, 1199–1259, WITH SPECIAL REFERENCE TO THE TOWNS. By Frank Burr Marsh. Pp. xi + 178. Cloth. $1.25.

VOL. III. THE COLOR LINE IN OHIO; A HISTORY OF RACE PREJUDICE IN A TYPICAL NORTHERN STATE. By Frank Uriah Quillan. Pp. xvi + 178. Cloth. $1.50. _____

CATALOGUE OF THE STEARNS COLLECTION OF MUSICAL INSTRUMENTS (Second edition). By Albert A. Stanley. With 40 plates. Pp. 276. $4.00.

LEGAL FOUNDATIONS
OF CAPITALISM

BY

JOHN R. COMMONS
University of Wisconsin

A^N investigation of the meanings of Reasonable Value as interpreted by the Courts, and consequently a contribution to the sciences of Economics, Ethics, Psychology, Law, and Politics. The book is an attempted correlation of these sciences into a volitional theory of value, and, as such, is a discussion of the foundations of Sociology in so far as applicable to an understanding of the modern capitalistic organization of Society.

This study of the origin and development of the legal precedents out of which present interpretations of law in America and England have grown is brought down to specific cases, covering almost every phase of economic life. It sets forth the theoretical basis on which judicial precedents are founded governing the conduct of economic life.

SUMMARY OF CONTENTS

8vo. 394 pages. $3.00.

THE MACMILLAN COMPANY

Publishers   64–66 Fifth Avenue   New York

A POLITICAL AND SOCIAL HISTORY OF MODERN EUROPE

BY

CARLTON J. H. HAYES

A New Edition of Volume II: 1815–1924

A FASCINATING story of Europe in the Nineteenth Century and the momentous years of the first quarter of the Twentieth. Hayes paints in broad lines and brilliant colors the shifting of national power, the conflicts of economic and political forces, and those social and intellectual revolutions which have moulded the present world.

This notably successful text is now ready in an enlarged edition containing five new chapters on the period from 1914 to 1924. The author summarizes the important events of the Great War, its significant diplomatic background, and the political, social, and economic aspects of the post-war developments.

The outstanding college text for history courses, on Modern Europe. Eight new maps are a feature of the revision.

CONTENTS OF THE NEW PART VI
"STORM AND STRESS"

8vo.   905 pages.   Maps. $4.00.

THE MACMILLAN COMPANY

Publishers   64–66 Fifth Avenue   New York

www.ingramcontent.com/pod-product-compliance
Lightning Source LLC
Chambersburg PA
CBHW030639270326
41929CB00007B/136